Improving the Efficiency of R&D and the Market Diffusion of Energy Technologies

Eberhard Jochem (Ed.)

Improving the Efficiency of R&D and the Market Diffusion of Energy Technologies

With Contributions by

Harald Bradke
Frank Marscheider-Weidemann
Oliver Som
Wilhelm Mannsbart
Clemens Cremer
Carsten Dreher
Jakob Edler
Bernd Ebersberger
Peter Radgen
Stefan Ruhland
Alexandra Krebs

Physica-Verlag
A Springer Company

Prof. Dr. Eberhard Jochem
Fraunhofer-Institut ISI
Breslauer Straße 48
76229 Karlsruhe
eberhard.jochem@isi.fraunhofer.de

ISBN 978-3-7908-2153-6 e-ISBN 978-3-7908-2154-3
DOI 10.1007/978-3-7908-2154-3
Springer Dordrecht Heidelberg London New York

Library of Congress Control Number: 2009920202

Cover design: WMXDesign GmbH, Heidelberg

Printed on acid-free paper

Springer is part of Springer Science+Business Media (www.springer.com)

Preface

Mankind faces several critical challenges associated with increasing economic growth, but also with the perspective of a longer and healthier life. To meet these challenges such as limiting climate change or developing new non-fossil fuel energy resources, research and innovation are the focus of interest and high expectations. However, research is expensive and the financial resources for research and development (R&D) are limited, particularly in emerging and developing countries. The risk of failure along the path from a good idea to market entrance and successful market diffusion is extremely high.

Given the high expectations of R&D for meeting the challenges on the one hand, and the limited resources and high risks on the other hand, the question arises whether the process of research and development can be made more efficient. Is it possible to achieve a given research and innovation target with fewer financial resources? Could some of the failures on the route from research idea to market penetration be avoided? Could research policies on the part of governments and companies use the same level of financial resources to accelerate innovation processes? There is a need to accelerate the introduction and diffusion of the new generation of technologies which are capable of meeting the described challenges and expectations.

This book reports on efforts to develop a methodology to use R&D budgets more effectively by identifying more clearly the status of the new technology, by setting clear research targets based on the technical and cost data of traditional competing technologies, and by identifying the important players in the innovation process at an early stage. The methodology is applied to four energy technologies, but the authors would like to stress that the methodology developed here could be applied to any research and innovation process that develops new technologies. Of course, the authors are aware that this methodology still has potential for improvement with ongoing progress in methods and data availability. The authors are very grateful to the German Federal Ministry of Economics and Technology (BMWi) for their trust and for providing the financial resources to investigate the question of whether the efficiency of innovations can be improved.

Contents

List of Figures

List of Tables

Contributors

Harald Bradke Fh-Institute for Systems and Innovation Research (ISI), Karlsruhe, Germany

Clemens Cremer EnBW Trading GmbH, Karlsruhe, Germany

Carsten Dreher Center for Cluster Development, Freie Universität Berlin, Berlin, Germany

Bernd Ebersberger MCI Management Center, Innsbruck, Austria

Jakob Edler Manchester Institute of Innovation Research MBS, MBS University of Manchester, Manchester, UK

Eberhard Jochem Fh-Institute for Systems and Innovation Research (ISI), Karlsruhe, Germany

Alexandra Krebs Hilti GmbH, Schaan, Liechtenstein

Willi Mannsbart Fh-Institute for Systems and Innovation Research (ISI), Karlsruhe, Germany

Frank Marscheider-Weidemann Fh-Institute for Systems and Innovation Research (ISI), Karlsruhe, Germany

Jürgen von Oertzen Fh-Institute for Systems and Innovation Research (ISI), Karlsruhe, Germany

Peter Radgen E.ON Energie AG, Munich, Germany

Frietsch Rainer Fh-Institute for Systems and Innovation Research (ISI), Karlsruhe, Germany

Sascha Ruhland MKMU Forschung Austria, Vienna, Austria

Oliver Som Fh-Institute for Systems and Innovation Research (ISI), Karlsruhe, Germany

Chapter 1
Executive Summary

Both industrialised and emerging countries face three major energy-related challenges in this century and over the next decades in particular: (1) The *share of fossil fuels in current primary energy use*, amounting to 80% globally, is likely to remain high during the next decades. This situation is in conflict with the pressing need to reduce energy-related CO_2 emissions which are driving global climate change – *a major threat* to mankind in this century. (2) The *peaking of oil production* which is expected *within the period 2015–2030* will cause sharp increases in fossil fuel prices if oil continues to increase. (3) The risks associated with sudden oil price increases are not small given the fact that *oil production will be re-concentrated in the Middle East* where two thirds of the remaining resources of conventional oil are located. The transportation sector is at particular risk as almost a hundred percent of the world's road, ship and air transport depends on oil.

Regarding the German energy situation, the structural aspects and risks are quite similar to the global picture: high per capita CO_2 emissions, particularly due to substantial coal use in electricity generation and domestic lignite resources, shrinking oil production in Germany and the North Sea, increasing dependence on oil imports from Russia and the Middle East, and a transport system which is largely dependent on oil products except for railways and trams.

This is why energy R&D and innovation policy will have to concentrate on accelerating innovation processes and the market penetration of those technologies capable of meeting these challenges in the future and reducing the risks mentioned above.

The lack of financing available to substantially enlarge public funds for energy-related R&D and market diffusion policies is a major bottleneck in meeting these challenges. One way to tackle this dilemma is to improve the efficiency of energy R&D and relevant innovations including market entrance and diffusion. Unfortunately, little is understood about how to make the best choices from among the myriad research ideas and proposals in order to reduce the risks of R&D funding and to maximise the outcome of public (or private) R&D funds.

This was the starting point of a small research programme called EduaR&D (*E*nergy *d*ata and *A*nalysis of *R*esearch *and D*evelopment) initiated by the Federal Ministry of Economics and Technology (BMWi) with five analytical projects on energy systems and related research and development. Two of them focus on

E. Jochem (ed.), *Improving the Efficiency of R&D and the Market Diffusion of Energy Technologies*, DOI 10.1007/978-3-7908-2154-3_1, © Physica-Verlag Heidelberg 2009

methodological questions ("Decision criteria for efficient R&D policy strategies" and "Setting priorities by methods of innovation and technology cycle research") and three on concrete technological areas ("Systemic evaluation of new cogeneration technologies", "Multi-criteria technology assessment applied to electricity generation", "Highly insulated buildings and intelligent building management"). The results of the second methodological project are reported here. During the analysis of this project, several meetings were held among the participating research teams to exchange and discuss intermediate results.

The attempt to improve the efficiency of energy R&D has also been taken up in the United Kingdom (by the Chief Scientific Adviser's Energy Research Review Group of the Office of Science and Technology) and in the USA (by the National Research Council). Both institutions published first reports on this issue in 2005.

1.1 Objectives of the EduaR&D Project Carried Out by Fh-ISI

The project's objective was to develop a new method for identifying and prioritising important energy technologies and energy-relevant topics of project-oriented energy research funded by the Federal Ministry of Economics and Technology (BMWi). Federal R&D funds should be allocated more efficiently by identifying bottlenecks and detectable risks and by supporting the stakeholder dialogue with comprehensive information.

The proposed methodology is based on four pillars:

- the use of *assessment tools of energy systems and energy economics* (impact on national energy demand and security of energy supply, potentials of emission reduction, cost reductions, co-benefits).
- The *heuristic concept of an innovation system* identifying important actors and their networks for research and diffusion of the specific technology using patent analysis and bibliometrics, interviews and literature evaluation;
- applying the *concept of technology cycles and market diffusion* in order to assess the status of technology development (by patent analyses, interviews, and market analyses); and
- using existing knowledge about the factors of success and failures of R&D projects.

The proposed assessment scheme was designed to work with quantitative indicators wherever feasible which could be aggregated using multi-criteria methods, but qualitative elements also had to be used. Making R&D processes and innovations in the energy field more efficient is too complex an issue to derive from theory, management science or energy economics. A huge amount of tacit knowledge is required to make decisions about these complex issues.

The concept was applied to *four selected case studies* for which the Federal Ministry of Economics and Technology (BMWi) is responsible and which display very different characteristics with regard to size and state of development, i.e. three types of fuel cells, passive solar buildings, CO_2 capture and storage of large fossil power plants, and industrial furnaces. The four technologies to be assessed in detail were selected on the basis of their energy economics relevance in Germany and globally as well as on R&D policy-relevant criteria such as their position in the technology cycle, size, type of companies and research institutions involved, and perceived risks. From this application of the conceptual approach, the main final deliverable of the project was a *cookbook-type manual* aiming to provide guidelines on how to move through information collection and analysis as well as the decision process.

The assessment methodology, its interim results and the first tentative recommendations were discussed at an international workshop in February 2006 to which experts had been invited from research, academia, industry and administration in the various technical areas and policy fields.

1.2 The Conceptual Approach

After the decision has been made about which technological fields should be assessed, the analysis relies on three different perspectives (see Fig. 1.1):

- The *energy economic approach* lays out how the new energy technology analysed will deliver energy services more cost efficiently, what its environmental impacts are and how it improves the security of supply. The analysis is based on a time horizon of 10 or 30 years depending on the energy technology considered.

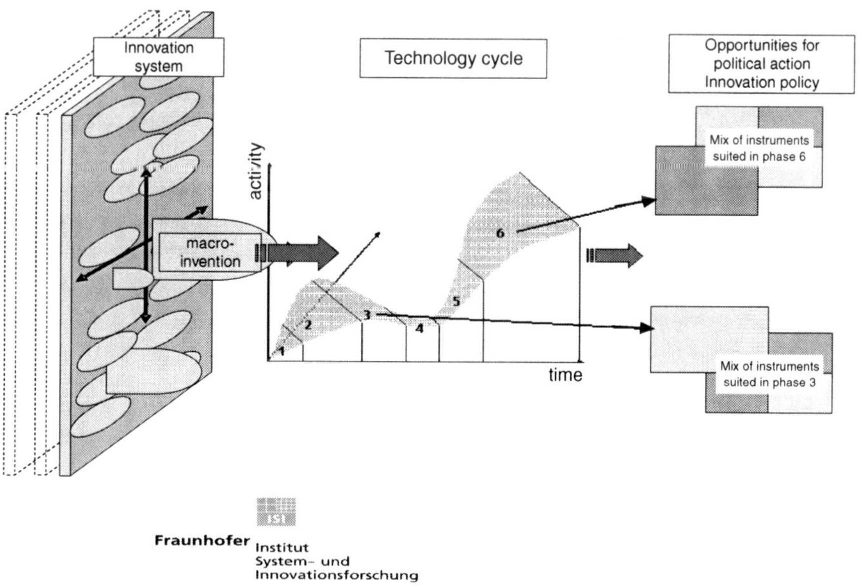

Fig. 1.1 Innovation system, technology cycle, policy options and their interplay in the three-method analysis suggested in this report. Source: Fh-ISI, 2006

- The *technology life cycle approach* investigates which phase the new technology is in. This step delivers essential information necessary to tailor the intervention measures to the specific requirements of the technology's phase. Where a plethora of new energy technologies is being considered for policy intervention, this step can identify promising technologies based on the technology cycle assessment.
- The *innovation systems approach* is based on the insight that innovations are generated and spread by the complex interaction of a multitude of actors. Innovation can only be understood by analysing the generation, dissemination and application of knowledge itself and the framework conditions for this process.

The outcome of the assessment integrating the three conceptual approaches (the energy technology and economics analysis, the technology cycle analysis and the innovation system analysis) has to answer several policy questions (see Table 1.1). The assessment process, therefore, followed these steps: (1) description of the technology and its potentials from the energy technology and economics point of view, (2) identification of position in the technology cycle and technical and financial bottlenecks of the technology (or elements), (3) analysis of the major players in the development of the innovation (4) policy recommendations for overcoming the bottlenecks identified and spurring the development (or market diffusion) of the technology considered.

In order to avoid misunderstandings of the assessment method developed here, it must be stressed that the *EduaR&D approach* should be regarded as

- an aid to reflect on the current energy research portfolio at the national or supranational level (i.e. EU or IEA), or at the corporate level of a large enterprise with several business areas in the energy field, or as
- *an assessment and policy design tool at the meso-level* of future energy technology areas and technology priorities.

It should not be seen as an aid for making funding decisions about concrete, individual research projects. Nor does the suggested approach replace basic political R&D decisions such as to not conduct applied research and development in the car industry or to prolong the research funding of high risk options such as energy fusion. The assessment is designed to be applied under specific circumstances (e.g. comparable risks involved) and specific conditions:

- after basic decisions have been made on how to *split the research budget into institutional and project funding* (e.g. relating to the budgets among the ministries (BMWi, BMU, and BMBF) at the federal level in Germany),
- after certain politically based decisions have been made at the macro-level on the further split of the project-based R&D budget (e.g. renewables versus efficient use of energy versus carbon capture and storage).

The assessment approach suggested relates to those *research areas and budgets which are programme- or project-oriented.* This approach is also useful for

Table 1.1 Integration of the three approaches in the EduaR&D process to answer policy questions

Funding questions	Energy technology and economics analysis	Technology cycle analysis	Innovation system analysis
What?	Determination of the energy economic significance of the new technology/ies, the necessary technological progress and cost level which could or should be achieved. Indications of barriers	Determination of the position in the technology cycle (patent analysis), definition of the technology at a "suitable" point, publication analyses (dynamics of the relevant knowledge production), market potentials	Determine which concrete level of the technology (module/product/system) and which system elements of this technology should be sponsored
Who?	Identify the energy users affected and technology attributes, the possible R&D funding institutions and possible research institutions	Assessment of the main actors as a result of the patent (and possibly publication) analysis, assess which types of actors (users, consumers) have to be approached	Identify the main actors in the respective subsystem and their needs and roles as well as possibilities of sponsorship leverage that may have an impact on research targets and specifications
How?	First indications of the prospective instruments for R&D or market diffusion (only tentative prior to the diffusion analysis and the innovation system approach)	The distribution of key knowledge holders and technology drivers may reveal possible gaps in cooperation. Market analyses make it possible to assess the ability of the market to absorb knowledge	Definition of the processes and activities which could/should be induced and reinforced via the sponsorship, i.e. via the conditioning of sponsorship instruments. Comprehensive approach results in large margins which may not fall under the portfolio of a specific ministry
Methodologies	Use of energy system analysis and projections, analyses of cost reduction potentials and environmental impacts	Patent and publication research; measured values for installations, units sold, projections of market diffusion	Modelling the subsystem, definition of the most important elements and then concrete actors. Definition of actor- and technology-specific questions, interviews

Source: Fh-ISI

decisions about changes in the R&D portfolio of highly funded energy technology areas on the one hand and less well funded ones on the other.

1.3 Results

First, it must be emphasised that this study only developed the methodology of the assessment; the concept still has to be applied in its full spectrum to individual technologies and research questions after completion of this study. This means that the results regarding the four technologies are likely to be incomplete and subject to further specification in the coming years. In particular, the analysis of the existing or future technological competitors of the studied technologies has not been fully applied in this analysis.

1.3.1 Analysis of Energy Technology and Energy Economics Aspects

The four selected energy technologies are highly relevant for a future sustainable energy system in Germany:

- *potentially avoided energy losses* range in categories between 100 PJ (industrial furnaces) and 300–1000 PJ (fuel cells and passive houses and buildings in particular);
- the *biggest CO_2 mitigation potential* ranges from 15 up to 50 Mt CO_2. Carbon capture and storage from fossil power plants would contribute the largest share, but at the expense of additional energy demand;
- raising the *security of energy supply by increasing the diversity of energy resources* would also be supported by carbon capture and storage technology (CCS), but also by highly efficient solutions that reduce imports of energy such as passive houses and buildings which also enable the inexpensive use of renewable energies;
- a *reduction in the costs of energy services* can be expected from passive houses (currently in the market entrance stage), industrial furnaces, and later possibly from fuel cells; as long as the externalities of greenhouse gas emissions are not included in energy prices (i.e. adaptation cost and remaining damage cost), CCS technologies will contribute to a higher cost of the energy system;
- the *longer term export potential* of all four technologies could be more than 3 billion € per year if they manage to successfully pass through the technology cycle and do so early enough compared to other competing industrial countries;
- the technologies also face *risks stemming from technical competitors* if clearly stated development goals (e.g. maintenance cost and periods, performance, and investment cost) cannot be reached; this is especially valid for fuels cells (vs. engine- and micro or mini turbine-driven cogeneration), passive houses (vs. low energy houses) and CCS where three technical options are being developed.

The US American Research Council also studied fuel cells and CCS technologies and, thirdly, efficient lighting as an example of a minor energy-consuming technology which nevertheless has extensive and very profitable efficiency potentials. The UK Report of the Office of Science and Technology focused almost exclusively on energy supply options (CCS, hydrogen, nuclear fission and fusion, photovoltaics, wave and tidal power). The 17 energy efficiency options mentioned in the report were only discussed superficially (including fuel cells, low energy buildings, and lighting).

The demand for greater differentiation in this study also requires greater data differentiation and data collection which is particularly challenging in the case of future projections. There is a trade-off between greater differentiation in the scoring per criterion and in the type of technology on the one hand and the additional efforts needed to calculate or collect differentiated data for a more disaggregated scoring and assessment on the other. Without a doubt, in many cases the improvement potential of the technological competitor has been (and still is) insufficiently considered in public R&D policy decisions.

1.3.2 General Conclusions on the Developed Method

The major lessons learned from the interplay of the three approaches and from their individual application within ISI-EduaR&D can be summarised as follows:

- *Methodological fit*: the design and the combination of the three approaches must fit the technological area considered and its identified elements. In-depth knowledge of both the technology and the context in the three analytical areas (energy technology and economics, technology cycle, innovation systems) is indispensable prior to designing the empirical work. This prerequisite may not be easy to meet in the immediate future within existing energy research administrations given the interdisciplinary character of this concept.
- *Procedure of the analysis*: the (partial) assessments made by each of the three analytical approaches should be carried out fairly simultaneously in order to make the analysis efficient and to enable cross-fertilization between the different methodological standpoints (see Table 1.1).
- *Analytical interaction:* target-oriented and quality assuring routines must be set up in order to secure the necessary feedback loops between the three approaches and the analysts and to secure the precision needed for the identification and assessment of the next steps in R&D, market introduction, or market diffusion policies. Ideally, a process manager with basic knowledge and experience in each of the three analytical approaches should be nominated for this task.
- The *level of technology and critical technical elements should be identified* as early as possible, particularly those representing the critical steps in the R&D process ("bottlenecks" e.g. acceptable maintenance intervals) or the market introduction process (e.g. acceptable cost or price levels of the new technology).

- Assessing the present and future advantages/disadvantages of *complementary or competing technologies* is an important element of the assessment as these may support or hinder the market introduction of the new energy technology in question (e.g. the PEMFC has to compete in stationary applications against cogeneration powered by the internal combustion engine or by Stirling engines, and the MCFC or the SOFC against micro-turbines or large cogeneration plants).
- The identification of *the relevant actors within the innovation system* is extremely useful, both for analysing the performance of the technological system and for getting the expert support needed for the various methodological steps. The analysis may start by patent or bibliometric analysis if key actors have limited knowledge, or by interviews if key actors are well informed, or when patent class identification is difficult.
- Moreover, how a technology fits into the *existing or future export profile* and thus matches the underlying economic structures of the country is one overarching variable for the funding decision. The ISI-EduaR&D assessment approach can also assist in determining how the specific technologies analysed relate to the existing technological strengths of the country in the context of global technological competition.

The assessment process suggested here runs the risk of consuming substantial resources in order to quantify all the indicators suggested (e.g. patent and citation development, anticipated costs at market introduction, estimated cost-cutting potentials by learning and economies of scale, present market shares). The trade-off between the additional efforts relative to the present decision-making process and additional benefits in terms of greater R&D efficiency will have to be decided case by case.

1.3.3 Specific Conclusions and Recommendations for the Four Technologies

1.3.3.1 Fuel Cells

PEM and SOFC fuel cells are still in the disillusionment phase or early re-orientation, whereas the MCFC is closer to market introduction. The DMFC has found some niche markets. All fuel cell types suffer from high investment costs due to the small production numbers projected and competing technical alternatives (engine or micro-turbine driven cogeneration) which are often overlooked when setting targets.

The discrepancy between the technological bottlenecks of the PEM and SOFC and the financial risks involved in mass production on the one hand and their underestimation by stakeholders on the other also calls for *stronger links between research funding* and *application-oriented activities* and the establishment of a fuel cell strategy.

Although the technology of DMFC is close to market introduction (present niche market in caravans and boats), R&D is still required to lower costs and improve the periphery systems and reliability.

Given the high level of international interest in this technology, German policy should aim to co-ordinate a German position on fuel cell priorities in the 7th Framework Programme of the EU more intensively and visibly.[1] This process requires the formulation of a concrete vision and technical targets for Germany involving the major stakeholders in a transparent process. The coordination of research and innovation funding should also be improved between ministries at the level of the individual German states.

1.3.3.2 Passive Houses and Buildings

In the rise phase of the technology cycle, passive houses and buildings face financing and cost issues as well as aesthetic issues.

The willingness to pay for higher initial building costs has to be increased by emphasising the lower lifecycle cost of passive houses and buildings. Appropriate financing subsidies and interest-reduced loans including no payback in the first years should be (and are) in place to bridge the financial gap and to further reduce the initial investment cost by learning and economy of scale effects.

Public procurement of passive buildings could have two benefits. First, it could increase the number of units demanded and hence induce scale effects in the supply of passive house components. Second, through a strict bidding process, it could reduce the profits of passive house suppliers and their component suppliers theoretically to an almost competitive level.

Financial incentives are essential for passive house suppliers, the suppliers of passive house components, pre-fabricated houses and system integrators to foster the development of and the demand for passive houses.

Due to its high technical standards, the passive house approach is driving the further development of components and promoting the development of new construction design. Since this is also applied in building codes (Energieeinsparverordnung EnEV[2]), new buildings and building refurbishments, the passive house approach is taking on the important role of a technology driver which should not be underestimated.

1.3.3.3 Carbon Capture and Storage from Fossil Fuel-Based Power Plants (CCS)

This technology is seen as being in the technology cycle phase between disillusionment and re-orientation. It has received a huge amount of attention recently

---

[1] A corresponding statement of the BMWi/BMBF was presented to the EU Commission on the 3rd of February 2005.

[2] Energy Conservation Ordinance

due to the pressing need to mitigate CO_2 emissions from large central energy conversion plants.

It is vital that the pilot and demonstration plants announced in Germany and elsewhere operate reliably in order to avoid severe disillusionment. Technical R&D activities which focus on specific bottlenecks are of major importance, e.g. inexpensive production of oxygen, safe operability of the combustion of fossil fuels under the high temperatures of oxygen-rich conditions or the alternatively envisaged recirculation of the flue gas. Another important issue is the analysis of the cost reduction potential of the three technical carbon capture alternatives depending on the time available, capital intensity and prices of emission certificate schemes.

Long-term tests are required to gain the necessary level of certitude about the leak tightness of geological storage. Beyond the exploration of storage sites, sufficiently accurate monitoring techniques are required for the CO_2 stored. Decisions on storage activities have to be prepared by investigating the current legal and societal conditions for CO_2 capture and storage and developing solutions for a legal framework and for public outreach.

1.3.3.4 Industrial Furnaces

Industrial furnaces are clearly at the diffusion end of the technology cycle. The technology is characterised by incremental innovations in material science and in better thermodynamic knowledge and control technologies. Given the high international competition in this field and the (still) excellent position of German manufacturers of industrial furnaces, the potential incremental innovation should not be underestimated, but rather supported by several actions.

The transfer of research results from power generation to furnaces is still taking too long or not taking place at all. One should consider linking these two research fields more closely in joint projects. The isolation of German research on thermoprocesses must be overcome using joint research projects and a greater openness of national funding and support actions for actors from abroad by increased cooperation in the European Framework Programmes.

The lack of cooperation regarding furnace technologies within the German innovation system could be overcome by funding networking, cooperative R&D projects, and interdisciplinary research groups. Improving burners and developing inexpensive but reliable sensors to analyse gas composition, temperature and flow velocity in harsh environments are still challenging research issues.

The German Government and the European Commission may also reconsider the R&D concept of SMEs and larger companies. International mergers – some of which are not pursuing new technologies, but rather additional profits as a short-term priority – are changing the structure of the manufacturers in this branch. As a consequence, SMEs are formally excluded from R&D incentives without the larger companies using the financial flexibility gained through the mergers to invest in technology progress.

1.3.4 The Manual – The Product of this Project

The authors of this report compiled a *manual on how to most effectively apply the proposed assessment concept* (see Annex I). It is expected that this manual will help to keep efforts and resources for an analytical endeavour of the suggested type at an acceptable level, given the additional benefits which are expected during the R&D phase and market penetration. The major rationale behind this is that bottlenecks and market obstacles will be identified very early on and the role of competing technologies will be analysed with greater intensity and objectivity.

1.4 Outlook

This proposed methodology made operational in the manual provides both experienced and younger energy research administrators with a quantified and transparent set of indicators allowing additional insights into the innovation system, the obvious bottlenecks and the next action to be taken, and makes their arguments more convincing and open for a rational discussion with stakeholders.

There is a good chance that the manual derived from the four case studies and the international workshop (see Annex) will contribute to improving the efficiency of energy research and diffusion policy in Germany in the near future.

The ISI research team suggests a next step to further improve and apply the new concept for R&D and innovation policy decision-making by the German Government. This next step should close an analytical gap that had to be left open due to the preliminary status of the concept's application. In particular, competing technologies should be included on an equal footing. Given the current activities regarding the design of a new energy programme of the German Government within the next 12 months, there may be a good chance to apply the methodology to some technological innovations such as fuel cell or carbon capture technologies regarding R&D and passive house regarding diffusion policy.

As the proposed methodology is generic in nature, it can also be applied (with some adaptation) to other important fields of new technologies such as health and medical treatment, food processing, or new materials.

Chapter 2
Introduction and Objectives of the EduaR&D Project Carried Out by Fh-ISI

Energy plays a central role in the economy of both industrialised and emerging countries. Every country is facing three major energy-related challenges in this century and over the next decades especially:

(1) The *share of fossil fuels in current primary energy use*, amounting to 80% globally and 78% in Europe, is likely to remain high during the next decades given the economics and limited acceptance of nuclear power and the (still) small economic potentials and market shares of renewable energies today and in the near future. This situation conflicts with the pressing need to reduce energy-related CO_2 emissions which are driving global climate change. These emissions cannot be sufficiently absorbed by the geosphere and have thus increased the atmospheric CO_2 concentration by 100 ppm or 35% since 1870. The *impacts of climate change are a major threat* to mankind in this century (IPCC, 2001, 2007).

(2) Recognising the role of crude oil as an energy price setter on world markets, energy policy will have to pay more attention to the *peaking of oil production* which is expected *within the period 2015–2030* depending on many influences such as global economic growth, technical progress and early investment in oil exploration and production, substitution by natural gas and renewable energies or more efficient energy use. With declining oil production, *energy price levels are likely to increase substantially*. Within this context, the alternatives mentioned as well as CO_2 capture and storage from centralised fossil fuel plants will receive increasing attention as possible backstop technologies and as a way of reducing the pressure on oil demand and the risk of high energy price increases.

(3) The increase in the price of fossil fuels is likely to be reinforced by the fact that the *oil production will be re-concentrated in the Middle East* where two thirds of the remaining resources of conventional oil are located. Given the present political instability in many countries of this region (often monarchies or dictatorships), there is some risk of oil price fluctuations in the case of longer supply disruptions. At particular risk is the transportation sector, as almost a hundred percent of the world's road, ship and air transport are dependent on oil.

E. Jochem (ed.), *Improving the Efficiency of R&D and the Market Diffusion of Energy Technologies*, DOI 10.1007/978-3-7908-2154-3_2, © Physica-Verlag Heidelberg 2009

This is why energy R&D and innovation policy will have to pay *more attention to speeding up the innovation processes and market penetration* of those technologies capable of meeting these challenges in the future and reducing the risks mentioned above.

However, the lack of financing available to substantially enlarge public funds for energy-related R&D and market diffusion policies is a major obstacle to meeting these challenges. One way to tackle this dilemma is to improve the efficiency of energy R&D and relevant innovations including market entrance and diffusion. However, little is understood about how to wisely select from among the myriad research ideas and proposals in order to reduce the risks of R&D funding and to maximise the outcome of public (or private) R&D funds.

2.1 Objectives of the EduaR&D Project

The project's objective was to develop a new method for identifying and prioritising important energy technologies and energy-relevant topics of energy research funded by the Federal Ministry for Economics and Technology (BMWi). Federal R&D funds should be more efficiently allocated by identifying detectable risks and supporting the stakeholder dialogue with comprehensive information.

The proposed methodology is based on four pillars:

- the heuristic concept of an innovation system identifying important actors and their networks of research and diffusion of the particular technology; this is done by patent analysis and bibliometrics, interviews and literature evaluation;
- applying the concept of product cycles and market diffusion in order to assess the status of a technology development (by patent analyses, interviews, and market analyses);
- using existing knowledge about factors of success and failures of R&D projects and
- extensive use of assessment tools for energy systems and energy economics (cost reducing potentials by experience curves, co-benefits, impact on national energy demand, potentials of emission reduction and others), but also using the tacit knowledge of experts and scientists in the particular fields of technology in research institutions and research laboratories in academia and industry.

The proposed assessment scheme is designed to work with quantitative indicators wherever feasible which could be aggregated using multi-criteria methods, but qualitative elements will certainly also be used.

The concept is being applied to four different case studies for which the Federal Ministry of Economics and Technology is responsible and which display very different characteristics with regard to size and state of development, i.e. CO_2 capture and storage of large fossil power plants, several industrial furnaces, three types of fuel cells, and passive solar houses and buildings with their specific technical equipment. If quantification is difficult, semi-quantitative methods are being used (by scaling).

The selection of the technologies to be assessed in detail has been performed on the basis of R&D policy-relevant criteria such as position in the technology cycle, large or small technology, energy-economic relevance, type of company and research institution involved and perceived risk of the R&D project. The assessment methodology and its interim results and first tentative recommendations were to be discussed in an international workshop on February 21 and 22, 2006 where experts in research in academia, industry and administration from the various technical areas and policy fields had been invited.

On the basis of the assessment tool finalised after the 2006 workshop the identified case studies were evaluated using specific tools available at the Fh-ISI such as energy system models to perform sensitivity analyses, multi-criteria assessment to aggregate the figures of the indicators and statistical methods.

A final report including conclusions on the practicability of the assessment scheme and recommendations for further R&D activities should terminate the project (see this report).

Reference

IPCC Intergovernmental Panel on Climate Change (2007). *Climate Change 2007 – Mitigation of Climate Change*. Fourth Assessment Report. Cambridge: Cambridge University Press.

Chapter 3
The Conceptual Approach

First of all, an attempt is made to classify each of the technologies examined using basic criteria which are of considerable importance for the rest of the analysis and significantly determine the priorities set in the subsequent analysis. These characteristics include (see Fig. 3.1): the regional scale of the markets, the predominant market form of the technology producers and the intensity of competition due to other technology options.

CO_2 separation technologies in large fossil-fuelled installations (likely to be made-to-order) such as e.g. in coal-fired power stations, or fuel cells (only competitive as mass products), are cited as examples for *global markets*. In contrast, passive

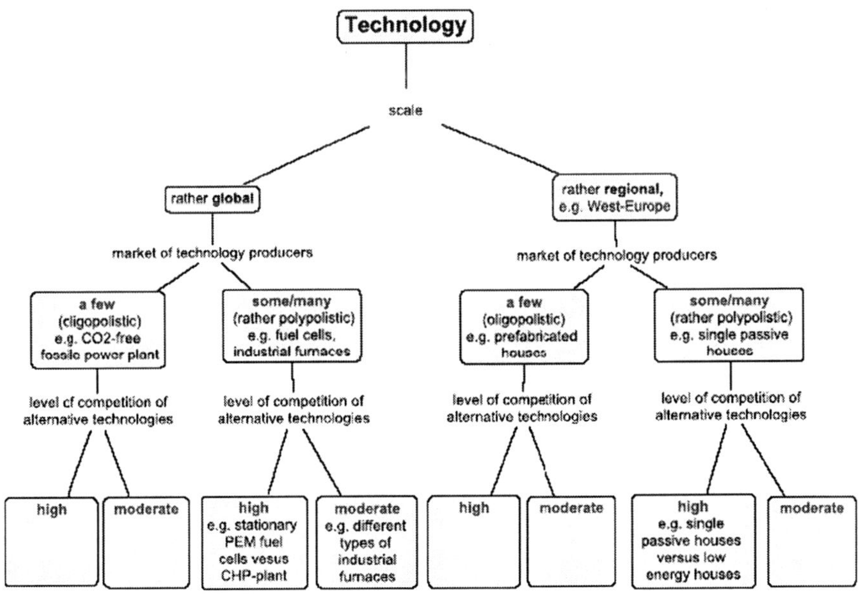

Fig. 3.1 Classification of energy technologies regarding regional scale, market type, and intensity of technological competition
Source: Fh-ISI, 2006.

E. Jochem (ed.), *Improving the Efficiency of R&D and the Market Diffusion of Energy Technologies*, DOI 10.1007/978-3-7908-2154-3_3, © Physica-Verlag Heidelberg 2009

houses represent a technology restricted as yet almost completely to German-speaking countries and Scandinavia. In the first case there are many global indicators which can be surveyed to obtain an accurate picture of the innovation systems and an appropriate degree of comprehensiveness of indicators, whereas the studies of regionally limited technology development have to comply with this regional restriction in many indicators (and may be limited as a result).

The *market form* for new technologies which do not require large infrastructures and do not necessarily have to be developed by very large enterprises is frequently polypolistic to start with but then tends to become more oligopolistic with increasing maturity of the technology (e.g. electricity generation, refineries, heat pumps, boilers). There are numerous reasons for this trend, but they are often linked with effects of economies of scale due to experience and the availability of capital to realize mass production. The market form at the point of introduction to the market has an influence on the policy measures selected at almost all stages of the technology cycle (certainly from the disillusionment phase onwards).

The *competitiveness intensity of an alternative technology* – whether this be a traditional one or one developing concurrently, with an as yet unresolved outcome of the technological competitiveness – has often not been sufficiently observed (e.g. heat pumps and conventional boilers in the late 1970s and early 1980s, the PEM fuel cell and the still improvable Otto engine today). If the technological competition obviously is (or could become) an open race, then these technological alternatives should be subjected to a similarly differentiated analysis in order to identify minimum development and performance targets of the new technologies considered and to correctly estimate their ability to penetrate the market or parallel markets.

3.1 Basic Idea: Three Angles of Analysis

In order to help raise the efficiency and effectiveness of policy intervention in the field of energy research, EduaR&D develops systematic information to guide future decisions on policy intervention (as well as energy research decisions in the private sector) aimed at fostering the generation and diffusion of new energy technologies (including energy-efficient technologies at the end-use and useful energy levels).[3]

The systematic collection and analysis of information on selected new energy technologies based on the approach(es) suggested in this document with the objective of being able to make better informed funding decisions is called the *"EduaR&D process"*.

The main final deliverable of the project was a cookbook-type manual describing the EduaR&D process; i.e. a document giving assistance on how to move through

[3] For brevity, where the use is not ambiguous we use the term 'technology' to refer to new energy conversion technologies or energy-related technologies at the useful energy level.

the information collection, analysis and decision process (see Annex 9.1). The project also demonstrates the essential parts of the manual by applying it (to a limited extent) to the four selected technologies.

This conceptual framework elaborates on the methodological approach. The EduaR&D process collects and analyses information from three distinct angles which are applied to the technology not sequentially, but simultaneously:[4]

- The **energy economics approach** lays out how the new energy technology analysed will (or may) deliver energy services or energy conversion more cost efficiently, what its environmental impacts are and how it improves the security of supply. The analysis is based on a time horizon of 15 or 30 years depending on the energy technology considered.
- The **technology life cycle approach**[5] investigates the production of the new energy technology based on patent and bibliometric analysis. The underlying idea is to identify in which life cycle phase the new technology is. This step delivers the essential information necessary to tailor the intervention measures to the specific requirements of the respective phase. Where a plethora of new energy technologies are being considered for policy intervention, this step can help to identify promising technologies, technical and cost bottlenecks of the new technologies considered, and related policies based on the life cycle assessment.
- The **innovation systems approach** is based on the insight that innovations are generated and spread by the complex interaction of a multitude of actors. Innovation can only be understood by analysing the generation, dissemination and application of knowledge itself and the framework conditions for this process. In the context of the EduaR&D process, the innovation systems approach makes a twofold contribution. First, it supports the delineation of the technology regarded and, second, it suggests possible interventions and actors by identifying factors which hamper the generation and diffusion of the technology considered.

Anticipating the discussion below, Exhibit 3.1 summarizes the six steps generating the systematic information needed to design energy R&D and innovation policy interventions. The analysis of this report also follows these steps which may be used in some iterations if new information or aspects are identified during the EduaR&D process.

---

[4]The simultaneous application of the three approaches causes the interdependence of the results since the results of one approach are fed back into all other approaches.

[5]Please note that the technology life cycle also includes the diffusion phase.

Exhibit 3.1 Steps to generate systematic information for energy policy design

1. *From an energy-economic and energy-policy perspective, identify those technologies which are likely to contribute especially to achieving energy policy objectives in the coming decades.*
2. *Determine the phase within the technology cycle for those technologies (and their components) eligible for promotion.*
3. *Identify the main technical difficulties, obstacles or market imperfections ("bottlenecks") of the selected technologies which determine the pace of development and currently make the innovation difficult or impossible.*
4. *Identify the competing technology and its potential for further improvement avoiding single technology focused R&D or innovation decisions and later surprises. Use this competing technology for setting R&D, performance and cost targets for the new technology.*
5. *Describe the respective innovation systems with the aim of locating actors and function correlations which are necessary or very useful in overcoming the identified bottlenecks, supporting the innovation process, and in identifying the improvement potentials of a competing technology.*
6. *Based on the knowledge of the innovation phase, the bottlenecks and the specific innovation system as well as the perspectives of the competing technology, identify suitable R&D measures or measures promoting diffusion.*

3.2 Selection of Energy Technologies Based on Energy Technology and Energy Economics Criteria

The selection of the energy technologies and their preliminary delineation is the key to achieving an ultimate improvement in the efficiency and effectiveness of policy intervention. The selection of the technologies has to be based on energy technology and economics criteria.

- What energy-economic benefits arise from the particular technology for Germany using a time horizon of 20–30 years (potentially up until 2050)?
- What is the probability of these benefits occurring in view of the open and highly uncertain future developments and framework conditions of the energy system of a country?

The benefits are evaluated based on a given set of energy policy targets. Supplementary, industrial policy considerations (e.g. existing or potential world markets) and economic issues (e.g. cost reduction potentials, possible mass production of the technology) and climate-related consequences (e.g. reduction of greenhouse gases and related adaptation costs) and their future dynamics are also employed as evaluation criteria for the energy technology.

Based on the set criteria summarized in Table 3.1, four technologies were selected to demonstrate the systematic information generation process:

- fuel cells in the three technological variants currently under development,
- passive houses and buildings, including their major technical aspects,
- CO_2 capture in fossil fuelled power plants and CO_2 storage in aquifers as well as
- industrial furnaces where specific types of materials (e.g. steel, cement, bricks, aluminium, glass) make this technological field differentiated and disperse.

This selection is also geared towards a set of heterogeneous energy technologies in order to be able to test the approach on a wide variety of conditions. The heterogeneity does not only refer to the technological configurations but also (presumably) to the respective technology life cycle phase of each technology, the regional aspects of the markets involved, and to both the shape and the configuration of the innovation system.

Ultimately, the information gathered by the EduaR&D process should enable a policy maker not only to decide on the policy options appropriate to fostering the generation and diffusion of a given technology, but also – under financial constraints – to select the next step or bottleneck of the technology progress or the most promising technologies from among a set of potential technologies. The process should also support changes of priorities in energy R&D and innovation policies in a transparent manner.

When these criteria were applied to a set of the four energy technologies selected, it became very clear that the quality of the information for the quantitative assessment varied among the technologies and that assessments with long-term future aspects become more difficult and uncertain if the technology under consideration is still in the early phases of development. This dilemma is obvious on the one hand in the case of CO_2 removal in fossil fuelled power generation where the technology is still dominated by R&D questions, and the smaller innovative steps in industrial furnaces, which are generally quite well known industrial processes, on the other. Given the limited resources for this analysis and its focus on the methodological issues, this selection assessment was made with little effort using existing literature; further in-depth analysis is advisable to receive a clear assessment on the basis of the criteria mentioned.

Given the uncertainties of the assessment without detailed analysis (see Sects. 3.2.1, 3.2.2, 3.2.3, and 3.2.4), a rather simple scoring was made to obtain a first insight into the suggested method and into relative importance of the four technologies from the energy technology and economics perspective. Each criterion in Table 3.1 received one to three scoring points, the results being as follows (see Table 3.2):

- The scoring does not differ greatly if only three categories are chosen; this is due to the fact that the avoided energy losses are highly correlated with avoided emissions and also to some extent with changes in specific energy system costs (the exception being CO_2 removal and storage).

Table 3.1 Selection criteria of the technologies examined from an energy-economic viewpoint

Selection criteria	Evaluation criterion	Characteristics, comments
Energy-economic benefit with regard to		
– Avoided energy losses up to 2030	> 300 PJ; 300–100 PJ; < 100 PJ	Applicable to efficiency solutions
– Avoided CO_2 emissions up to 2030	> 50 Mt; 50–15 Mt; < 15 Mt	By efficiency or substitution
– Other avoided important emissions	> 1 Mt; 1–0.3 Mt; < 0.3 Mt	
– Improved supply security up to 2030	σ (statistical distribution)	New energy source
– Cost efficiency after take-off phase	<–10%; –10 to + 20%; > 20%	Internal rate of return
– Change in the specific energy system costs in the field of application:	> 80%; 50 to 80%; < 50%	Indicator for market potential
The possibilities of the application with regard to		
Export potential for industry	> 3 bill. €/a; 1–3 bill. €/a; < 1 bill. €/a	Indicator for economy of scale potential
Regionally specific/globally standardized technology	Scores between 1 and 3	Large differences between technologies
Individual/mass product application	Scores between 1 and 3	Cost degression potential
Within the promotion competence of the Ministry of Economics	Yes/no	
Particular advantages over technical competitors [e.g. less maintenance, high co-benefits, etc.]	Scores between 1 and 3	Technical potential

Table 3.2 Assessment of the four selected technologies by the criteria of energy technology and economics

Selection criteria	Fuel cell	Passive houses and buildings	CO_2-free power generation	Industrial kilns and furnaces
Avoided energy losses	300–1000 PJ/a	300–1000 PJ/a	–300 PJ/a (–1)	< 300 PJ/a
Avoided CO_2 emissions	< 15 Mt/a	15–50 Mt/a	15–50 Mt/a	< 15 Mt/a
Other important avoided emissions	0.3–1 Mt/a	< 0.3 Mt/a	> 1 Mt/a	< 0.3 Mt/a
Important supply security up to 2030	Little (1)	Medium (2)	High (3)	Little (1)
Cost efficiency after market take off	One of the R&D-goals (1)	Yes (2)	Only under emission certificate regimes (2)	Yes (2)
Change in the specific energy system costs	Little reduction (1)	Eventually medium reduction (2)	Substantial increase (–2)	Little reduction (1)
Total scores	**8**	**11**	**7**	**7**

Source: Workshop presentation

- Passive houses and buildings have the best score (11 points) where new houses and buildings and some retrofit of the existing building stock has been assumed. The lowest scores (7 points) are for industrial furnaces (lower efficiency potentials) and CO_2 removal from power generation (adding to energy losses of conversion and to the energy system costs).
- From the technological and economic point of view, the uncertainty attached to the scoring is probably the highest for fuel cells, as these have either remaining technical bottlenecks to overcome (PEM, see Sect. 4.1), or steep cost reductions to achieve before they will be competitive with other technical options such as engine-driven cogeneration, micro-turbines, or Stirling engines, or with larger systems such as large cogeneration/heat pumps.

Looking at the other selection criteria (export and mass production potential, regionality, and advantages over technical competitors), the differentiation among the four technologies becomes more obvious. However, the total scoring shows very similar importance of the four technologies selected although they have different patterns of the scoring according to their specific properties (see Table 3.3):

- Properties of regional or global markets, the potential of mass production and related cost reduction potentials, as well as specific technical advantages become more influential in the scoring and put *fuel cells on top (19 points)*, while the winner based on energy-economic criteria, passive houses and buildings, falls back into the second position (18 points) due to lower export potentials, limited regionality, cost reduction potentials and limited advantage over its competitor (low energy houses and buildings).

Table 3.3 Assessment of the four selected technologies by several criteria (export potential, mass production potential, technical advantages)

Criteria with regard to application probability and ancillary benefits	Fuel cell	Passive houses	CO_2-free power generation	Industrial kilns
– export potential for industry	> 3 bill €/a	1–3 bill €/a	> 3 bill €/a	1–3 bill/a
– regionally specific (1)/globally standardised (3)	3	2	3	2
– individual (1)/mass product application (3)	2	1	1	2
– particular advantage over technical competitors	+++ (3) (engine driven cogeneration)	++ (2)	+ (1)	++ (2)
Subtotal	11	7	8	8
Total scores (including Table 3.2)	**19**	**18**	**15**	**15**

Source: workshop presentation.

- Although CO2-free power generation technologies receive eight points in this second assessment set, they equally obtain the third position in the scoring (15 points) due to their smaller importance from the perspective of energy economics.

To conclude, the differentiation has to be refined, and the weight given to each criterion has to be considered in more detail. The differences in the assessment within a technical field (e.g. fuel cells with three types or industrial furnaces for many different industrial branches) suggest it might be necessary to distinguish individual types of the technological areas in more detail.

However, the demand for greater differentiation also requires more data differentiation and collection which is particularly demanding in the case of future projections of the development of each technology considered. Thus there is a trade-off between more differentiation in the scoring per criterion and also in the type of technology on the one hand and the additional efforts needed to calculate or collect differentiated data for a more disaggregated scoring and assessment on the other. Although fuel cells had the best score in this ad hoc assessment, the authors are convinced on the basis of the results reported in Chaps. 4, 5, 6, and 7 that the set of criteria has to be enlarged and the scoring has to be differentiated.

3.2.1 Applied Boundary Conditions and Energy Economics Projections

Since the energy economics-related analysis could only rely on existing literature, two comprehensive reference and target-oriented scenarios describing possible future developments of the German energy system could be used as the major sources for the analysis; these two scenarios were published during the last few years (German Bundestag Enquête-Commission "Sustainable Energy Supply" (Nachhaltige Energieversorgung unter den Bedingungen der Globalisierung und Liberalisierung), 2002; and EWI/Prognos, 2005). Model calculations were carried out aiming at an 80% decrease of greenhouse gas emissions by 2050 (relative to 1990 emission). Data on framework conditions and energy drivers (e.g. population, economic growth, living and office floor areas, industrial production, energy price projections) that have been assumed or calculated for the period 2030–2050 differ to some extent between the two sources (see Table 3.4).

The economic growth after 2030, assumed to be 830 €/cap. and year by the Enquête Commission, is double Germany's growth over the last 50 years and has to be judged as quite optimistic. This leads to a higher energy demand than can be expected realistically. This difference in growth is also reflected by the drivers of energy demand for freight transport which assume a doubling of tonnes km within the next five decades.

The two studies share a similar view of the development of living floor area in private households and of passenger transport until 2030. There seems to be little

Table 3.4 Socio-economic boundary conditions and energy drivers of several energy demand projections, Germany 2030–2050

	2000	2010 EWI/ Prognos	2010 Enquête	2030 EWI Prognos	2030 Enquête	2050 Enquête
Population (mill.)	82.2	82.4	82.1	79.3	77.9	67.8
Gross Domestic Product		**2000/10**	**2000/10**	**2020/30**	**2020/30**	**2030/50**
Change/year (%)		1.3%	1.9%	1.3%	1.3%	**1.0%**
GDP, Bill. $Euro_{1995}$	1,970	2,238	2,438	2,960	3.286	3.990
Δ GDP/cap., Euro/cap.a		325	508	520	650	830
Labour statistics (mill.)	**2000**	**2010**	**2010**	**2030**	**2030**	**2050**
Working population	37.5	38.9	37.6	37.5	34.9	32.2
Unemployed persons	4.4	n.a.	5.0	n.a.	2.0	1.7
Private households (mill.)	38.2	39.7	n.a.	39.7	n.a.	n.a.
Living floor area						
Total (mill. m^2)	3,308	3,615	3,733	4,406	4,231	3,972
m^2 per capita	40.2	43.9	45.5	55.6	54.3	58.6
Transport performance						
Passenger transport (Bill. person km)	968.1	1,110	1,090.7	1,060	1,139.1	1,026.9
Freight Transport (Bill. tonne km)	483.1	589	607	785	839	964
Energy prices (€/GJ)						
Oil (cif)	2.81	2.50	3.56	3.35	5.06	6.57
Natural gas (cif)	2.15	2.17	2.84	2.80	4.20	5.57
Hard coal (cif)	1.36	1.44	1.43	1.58	1.76	2.09
CO_2 emissions Reference Sc. (Mt)	853	837	825	717	816	700

Sources: Enquête-Commission, 2002; EWI/Prognos, 2005.

interrelationship between long-term economic growth and the prices of imported energies, as the Enquête Commission (2002) expects higher economic growth and higher energy prices compared to EWI/Prognos (2005; see Table 3.4).

Interestingly, neither projection makes clear assumptions on the application of fuel cells and passive houses or on the efficiency gains of industrial furnaces. In the case of fuel cells, the competition among different technical options is left open (e.g. internal combustion engine, Stirling engine, micro-turbine, and fuel cell) and in the case of passive houses, the alternative of low energy houses and buildings is not explicitly discussed. Regarding CO_2 capture and storage, quantitative assumptions are made in both studies ranging from avoiding 35 Mt CO_2 (EWI/Prognos, 2005) to 75 Mt CO_2 in 2035 (Enquête, 2002).

3.2.2 Decentralized Electricity and Heat Generation – Fuel Cell Options and Their Competitors

The intensified use of small combined heat and power (CHP) generation units (engine driven and micro-turbines) as well as new technologies such as fuel cells in decentralized power supply are receiving more attention due to their capability to reduce CO_2 emissions in the future. On the other hand, reduced, long-term space heat demand and a growing market penetration of renewable energy sources in the future are limiting factors for the potential of CHP and thus also for fuel cells.

Presently, decentralised micro CHP plants using gas engines or gas turbines are becoming increasingly market relevant. The use of fuels from RES is still a long-term goal. Electrical capacity is estimated to be more than 5,500 MW, of which 2,000 MW is motor-driven and 3,500 MW generated using gas turbines. The power spectrum of these plants varies from a few kW up to the MW range. The Enquête-Commission (2002) argues that increased electricity generation at constant heat delivery and development of the following potentials are strategies to make progress in this field:

- decentralised district heat grids ("Nahwärmenetze"), provision of infrastructure for fuel cells provided that technical and cost targets are achieved; (Pehnt & Traube, 2004)) or
- house and building supply systems with small capacities ("Objektversorgung").

For cost reasons, the Enquête-Commission and EWI/Prognos assume that new large district heat grids or extensions are not likely in the future. The estimates of the technical potentials for (large and small) CHP range between 220 TWh and 380 TWh per year (today, about 50 TWh per year). Around 50% of the above-mentioned upper potential depends on the availability and chances of efficient CHP plants in decentralised object related supply and buildings supply with micro CHP and fuel cells.

Table 3.5 Future CHP electricity generation in Germany (net TWh): small-scale CHP and stationary fuel cells by 2010–2030

	1998	2010	2020	2030
Sustainable scenario 1 (WI) "Umwandlungseffizienz"*	0	1.9	20.6	43.1
Sustainable scenario 2 (WI) "REN-REG-Offensive"**	0	1.9	26.9	35.6
Scenario "Fossil-nuclear" (WI)	0	0.2	1.6	3.7

* Conversion efficiency.
** Renewables/rational energy use offensive.
Source: Enquête-Commission, 2002.

The Enquête-Commission expects an increase in electricity production of between 36 and 43 TWh in 2030 representing about 9% or 8% of the expected total net electricity use (see Table 3.5).

In the sector of micro-CHP as well as for CHP installations in the 100 kW–2 MW class, there are high expectations of stationary fuel cells due to their high efficiency rates (and the higher ratios of electrical to heat output also at partial load), low maintenance costs, very low to zero emissions and reliable power control. However, neither projection answers the question of how much of this potential is likely to be realised using fuel cell technology. This is a clear hint that the *technological competition* among today's micro CHP technologies, the internal combustion engine and the micro-turbine is an open race which is likely to become even more competitive due to the introduction of the Stirling engine in small and lower temperature applications such as buildings within the next few years.

3.2.3 Low-Energy and Passive Houses and Buildings

Buildings are termed "low energy buildings" if their energy demand per floor area is below the existing building code by more than 30%. If the energy demand is even lower than this, then the following terms are applied depending on the originator or region: "Ultra-Niedrigenergiehäuser" (ultra low energy houses, Gertis), "Plus-Häuser" (plus energy houses, Disch), "Quasi-Nullenergiehäuser" (quasi zero energy houses, Humm), "Passivhäuser" (Passive houses, Feist, 2005) or "Minergiehäuser" in Switzerland. These (mostly newly built) houses or buildings manage to produce a comfortable in-door temperature without an active heating or air conditioning system in moderate climatic regions of Europe (Feist, 2005). A specific energy demand of less than 15 kWh/m^2.a is presupposed as a central criterion.

Low-energy buildings constructed in recent years show that a specific space heat demand of around 30–70 kWh/m^2.a does not induce noticeable increases in construction costs. The result of the German Bundestag Enquête-Commission (2002) assessment was that a decrease in specific costs of heat energy by 50% is actually possible due to learning and economy of scale effects. Since the introduction of the current building code (Energy Conservation Ordinance, EnEV) in 2002, the standard for new houses and buildings comes close to low-energy buildings.

A study by Kleemann (2000), sponsored by the German Bundestag Enquête-Commission, investigated the future energy efficiency potential in the German building stock (considering refurbishment activities but also demolition of buildings). For residential space heating, energy saving rates of 28% or 570 PJ were ascertained by 2020 (2050: 68% or 1420 PJ). The assumptions made by this survey included intensified standards (low-energy standard from now on and low-energy standard+ from 2020 on), and realising the full efficiency potential in refurbished houses and buildings.

The 2005 EWI/Prognos Energy Report IV has no detailed information on residential space heat with regard to passive buildings. Therefore, the focus is on the Enquête Commission Reference scenario (2002) (see Table 3.6).

- The standards for new buildings correspond fully to the Energy Saving Ordinance EnEV 2002, under which requisitions will be tightened at certain intervals resulting in values lower than today by 30% (2030) and by 40% (2050). For the Sustainable Scenarios 1 and 2, the requirements will be tightened or correspond at least to the reference case.
- There are mostly analogous requirements for older buildings (see Table 3.6; older buildings 2050: –50%), which presume, however, that as in the past only 20% of renovations include thermal insulation. In the Sustainable Scenarios, the renovation rate for older buildings was expected to increase by 30% compared to the reference case.

The model calculations showed the following results (basis 1998: final energy use residential space heating 2100 PJ, hot water preparation 250 PJ):

- Reference scenario: final energy for space heating and hot water is expected to decrease by 18% (1998–2050).
- Sustainable scenario 1 ("Umwandlungseffizienz"): the IER Institute assumes increased renovation intensity as follows: 40% by 2010, 60% by 2020, 80% by 2030 and 100% afterwards. Thereby potential efficiency improvements may initiate reductions in energy use for space heating and hot water by 37% in the period from 1998 to 2050. The WI Institute's renovation rates (1.5%/year) correspond to a complete overhaul of substantial components of buildings such as windows and facade/storefront of the building stock *within 65 years*. It is also postulated to ensure compliance with existing requirements for energy-renovation measures. The renovation of older buildings together with the construction of low-energy new buildings may decrease the energy demand for space heating and hot water by 33%.
- Sustainable scenario 2 ("REN-REG-Offensive"): model calculations by the IER also document substantial efficiency gains. The energy used for space heat/hot water is expected to decrease by 40% between 1998 and 2050. Moreover, GHG reductions will occur from substituting fossil fuels by electricity as well as

Table 3.6 Characteristics of new buildings and refurbishment of older buildings

	Time horizon	Reference scenario	Scenario Conversion Efficiency		Scenario "REN-/REG-Offensive"	
Minimum requirement new buildings	from 2002	corresp. EnEV	at least reference		at least reference	
	2020	EnEV – 15%	at least reference		at least reference	
	2030	EnEV – 30%	at least reference		at least reference	
	2050	EnEv – 40%	at least reference		at least reference	
Minimum requirement refurbishment of older buildings	from 2002	corresp. EnEV	at least reference		at least reference	
	2020	EnEV – 15%	at least reference		at least reference	
	2030	EnEV – 30%	at least reference		at least reference	
	2050	EnEv – 50%	at least reference		at least reference	
Renovation rate older buildings			Reference + 30%		reference + 30%	
Rate of implementation older buildings (% of stock)	by 2010	0.5%/a	WI	IER	WI	IER
	by 2020	0.5%/a	1.5%/a	1.0%/a	2.0%/a	1.0%/a
	by 2030	0.5%/a	1.5%/a	1.5%/a	2.0%/a	1.5%/a
	afterwards	0.5%/a	1.5%/a	2.0%/a	2.0%/a	2.0%/a
			1.5%/a	2.5%/a	2.0%/a	2.5%/a

WI = Wuppertal-Institut, IER = Institut für Energiewirtschaft und rationelle Energieanwendung Stuttgart University.
Source: Enquête-Commission, 2002.

> renewable energy sources. In addition, the market introduction of hydrogen-based fuel cells will be accelerated in the residential sector. The WI scenario assumptions are more or less the same as a complete refurbishment/renovation of the most relevant building components *within 50 years*. The technical progress achieved in the new buildings sector will be implemented gradually in older buildings as well. Over the course of time, the contribution of renewable energy sources will increase, especially via local heat networks and efficient centralised equipment instead of individual heating systems. Temporarily, the use of gas-based fuel cells will be intensified. Increased renovation rates with regard to the reference scenario will induce additional new products and innovations by the construction sector and the insulating materials industry. The calculations by IER show an energy consumption decrease of around 45% for the generation of space heat/hot water (Enquête-Commission, 2002).

Keeping in mind the demographic development according to the Reference scenario, the absolute residential energy demand for space heat/hot water will decrease by more than 400 PJ by 2050. With reference to this, improved energy efficiency (see Sustainable Scenarios) in new buildings (low energy buildings and passive energy buildings) will bring about a decrease of 800–1000 PJ. This potential corresponds to around 10% of the present total energy demand.

The results of other studies (e.g. Kleemann et al. 2000) are smaller than those of the Sustainable Scenarios 1 and 2: for new buildings and the period from 2005 to 2020, a reduction of 22% compared to the current building code and 75% thereafter results in savings of 24.5 PJ (2020) and 148 PJ (2050) respectively.

Comparing new buildings and the renovation of older buildings shows that concentrating on new buildings would only tap about 15% of the energy efficiency potential compared with a strategy combining the potentials of new buildings together with the renovation of existing buildings.

3.2.4 Carbon Capture and Storage from Coal-Based Power Plants

Mitigation options for stabilising greenhouse gas emissions include CO_2-free fossil fuel based power plants, i.e. CO_2 capture and sequestration (CCS). In this context, the security of the energy supply in OECD countries and the vast coal reserves in Asia are major drivers for developing this technology and for energy policy actions in the next decades.

Results on future global carbon dioxide emissions vary; they could be 60–90% higher by 2030 than the targets set for 2010 by the Kyoto protocol. A study by IEA (2004) assessing the global potential of CO_2 capture and storage pointed out that CCS might be feasible but that more pilots and demonstration projects are needed together with more basic R&D in order to improve the technology and integrate it into new combustion processes. In the field of fossil fuel power generation, it is estimated that it might be possible to reduce CO_2 emissions from large industrial

Table 3.7 Global profile by process or industrial activity of large stationary CO_2 sources (>0.1 Mt CO_2 per year)

Industrial processes	Number of large sources	Emissions, Mt CO_2/year
Power	4,942	10,539
Cement production	1,175	932
Refineries	638	798
Iron and steel industry	269	646
Petrochemical industry	470	379
Other	n.a.	83

Source: IPCC, 2005.

plants by 80–95% if the CO_2 is separated, compressed and then transported to a storage reservoir (Cremer, 2005).

In 2003, global emissions from fossil-fuel use amounted to 25 Gt CO_2 (IEA, 2005). More than 50% of this is attributed to large stationary emission sources (> 0.1 Mt CO_2/yr.), i. e. these sources are main candidates for CO_2 capture (IPCC, 2005; see Table 3.7). Adopting CCS would have significant consequences for changes in energy policy, including a long timeline of active support, more incentives such as "blame-free continued use of coal" (major source in the US, China and India; note: the IEA emphasised the need to help China and India to adopt the CCS technology and for higher investments to develop and deploy this technology on a large scale.

Other fields of activity include locating suitable geological formations for storage, removing barriers and reducing uncertainties (Cordis, 2005). If these activities are successful, CCS would then have the potential to save emissions for decades during the transition to non-fossil fuels. Besides technical feasibility and operating efficiency it is necessary to develop comprehensive general conditions on the legal side (environmental law, nature conservation, mining law and water rights). The introduction of the European Emissions Trading System (ETS) was a major prerequisite for developing the economic and legal boundary conditions for CCS.

For Germany, the CO_2 reduction targets after 2012, the post-Kyoto period, will be of major relevance for the power sector given the vast lignite resources in Germany. Recently, the construction of pilot plants for carbon dioxide free power stations has been announced by several large electricity suppliers. Two utilities are ready to invest in pilot CCS plants at locations in Germany:

- Vattenfall Europe will start with a 30 MW pilot plant (2008). If tests are successful, the construction of a 300 MW demonstration plant is planned until 2015 (Schwarze Pumpe/Lausitz). Commercial use will not be realisable until 2020.
- RWE Power will start operation of the first CO_2-free power plant (400–450 MW) in 2014 (incl. storage). According to BGR (Bundesanstalt für Geowissenschaften und Rohstoffe), the storage capacity in Germany is between 23 and 43 billion tons, i.e. if the mean figure of 33 billion tons is accurate, the deposits could

absorb the total CO_2 emissions of domestic power stations for a period of around 80 years (RWE, 2006).

In 2030, the CO_2 reduction potential of German coal and lignite based power plants is estimated to be on the order of 100–120 Mt of CO_2 (Enquête Commission, 2002, Scenario Conversion Efficiency). The Europe-wide replacement of outmoded power plants will provide the chance to introduce a new low-emission power plant generation. CCS may represent one conceivable bridging technology until renewable energy sources are sufficiently developed in view of the fact that the available storage capacity in Germany is limited (Radgen et al. 2005).

3.2.5 Industrial Furnaces

Efforts aimed at reducing greenhouse gas emissions focus especially on cross-cutting technologies. In Germany, about 70% of total final energy is used by these technologies which include industrial furnaces, steam and hot water generators up to 50 MW_{th}, drying processes and heat exchangers, small scale CHP systems, electrical drives, pumps, compressed air systems, ventilation and air-conditioning, refrigeration and lighting.

Industrial furnaces are using the largest share among these cross-cutting technologies; around 950 PJ/a are consumed in several industrial basic product branches (around 840 PJ or 85% by fossil fuels and 115 PJ or 15% by electricity (Fraunhofer-ISI, FfE, 2003, Radgen, 1998)). Application fields of industrial furnaces includes melting (particularly steel and non-ferrous metals), heating for casting and moulding purposes, heat treatment, sintering and calcining (see Sect. 7.1). In the commercial sector, furnaces are important in bakeries, glass and earthenware production and the metal industry. Although the related CO_2 emissions amount to 75–80 Mt CO_2 per year, the recent energy demand projections do not explicitly mention furnaces' efficiency, technology or fuel substitution potentials.

The *burner system* of furnaces using fossil fuels is one of the key components for further improvement (Schmid, 2003), e.g. high-speed burners, radiant burners (where heat transmission is performed by radiation), high temperature air preheat burners with low emissions, recuperative and regenerative burners and the flameless oxidation burner. Other important technological options to reduce energy demand are improved insulation materials, better process control and heat recovery (see Table 3.8).

Many of these measures are economical even at today's energy prices, but their implementation is often limited to new acquisitions or to retrofitting investments. Altogether, the technical energy efficiency of new generation furnaces is estimated to be up to 30% better than present installations. The total assessed economic energy saving potential of cross-cutting technology "industrial furnaces" is around 90 PJ (10%) at present in Germany. Further R&D could push this even further.

Table 3.8 Energy saving possibilities and options for industrial furnaces

Target	Measures
Reduction of heat loss and surface temperature of the furnace walls	Thermal insulation materials in the fire-resistant lining
Optimum process control	Model-based control
Heat recovery	Using the waste heat contained in the hot exhaust gas to preheat the combustion air and/or the fuel gas[1]
Reduction of exhaust gas loss and reduction of energy demand	Use of oxygen-enriched combustion air
Increased performance of the furnace	Principle of permeable radiation walls (energy saving of about 2–5%)

[1]The average fuel efficiency of industrial furnaces installed today is around 60%.
Sources: Schmid, 2003, Fraunhofer ISI, FfE, 2003.

3.3 Delineation of a Technology

Once the technology has been identified, it then has to be delineated. Which level of the technology, which of its variants, which of its components will be analysed using the EduaR&D process is finally co-determined by the targets of the potential policy intervention, since this intervention aims the greatest possible leverage on the development and diffusion of the technology. This leverage varies from level to level, variant to variant and component to component. The functional role of the different variants and the different components has to be taken into account as well.

For the analysis of the innovation system, the EduaR&D process needs a definition of the technology level at which the potential policy intervention will take place. Technology levels can be:

- a single technique (i.e. a certain bundle of knowledge promising benefits e.g. CO_2 separation via synthesis gas method),
- a major module for the technology (e.g. the gasifier of the PEM fuel cell),
- a product (e.g. the hydrogen turbine) or a
- complex techno-economic system integrating various modules and products (e.g. zero CO_2 electricity generation based on lignite including lignite extraction and CO_2 storage).

The starting point for the delineation of the technology is a detailed description of the technology, its variants, components, modules and techniques. The clear demarcation of which level of the technology will be analysed in the EduaR&D process is not possible in a single step. Instead, communication with the technology cycle approach and the innovation system approach in the shape of feedback loops facilitates the definition of the technology level iteratively. Essential elements or levels of the technology which must be included in the analysis are identified as technological and economic bottlenecks. Observations from the innovation system approach about *systemic bottlenecks* in the development and diffusion of *the technology* and particular *opportunities for the technology* will also play a central role in reshaping the definition of the technology for the analysis. Tractability and feasibility issues also have to be considered.

No fixed set of finite rules can be presented for the definition process of a technology in the EduaR&D process (see Exhibit 3.2 and the example of the PEM fuel cell).

Exhibit 3.2 Example for the delineation of a technology

In all four technology areas it was very difficult to delineate the technology to be examined (i.e. determine the level) as well as to select possible variants.
For example, for the *fuel cell*, an early decision was made in favour of PEM due to the evident dynamics in this field and the early insight that the actor constellations in the three fuel cell variants are completely different. The basic technological idea is the only shared result of the analyses in this field. Furthermore, within the PEM, two obvious weak points were identified at an early stage of the project: its short lifetime and the costs of the PEM membrane. The analysis then attempted to look at all three variants – because of their potentially large energy and industrial policy significance – and now faces the challenge of having to describe three innovation systems which are largely non-overlapping.

3.4 The Analysis of the Technology Cycle

3.4.1 General Model

The development of technologies is shaped, hindered or promoted by the demand of users and by socio-technological paradigms (Dosi, 1988) leading to the development of a basic technology in certain trajectories. At the company level, technology management has to consider the proper time for switching a technology or staying within the current path (Christensen, 1997), or whether the diffusion potential into other applications makes it worthwhile to stick with a certain technology (Gold, 1981). At the policy level, the promotion as well as the promotion instruments and arrangements within the innovation system have to be considered and adjusted (Edquist, 1997, cf. e.g. policy instruments timing based on diffusion metrics, Dreher, 1997).

In order to visualise and condense different technology dynamics approaches, Meyer-Krahmer and Dreher (2004) presented a descriptive model for a basic technology cycle combining the science base cycle with the innovation diffusion model (see Fig. 3.2 and Dreher et al. 2005).

After the first stage of discovery and exploration, disciplinary and interdisciplinary research investigates the opportunities for the new technological principals. This leads to euphoria about the new technological possibilities among the growing community of scientists and applied researchers (stage 2). Over time, however, several options turn out to be either technically or economically unfeasible. Therefore, research activity in these areas is reduced or stops altogether (stage 3). Because

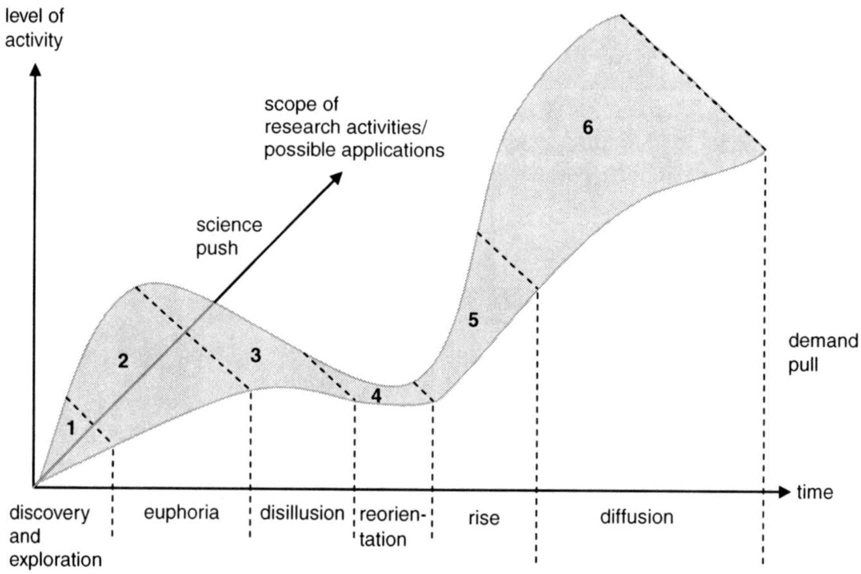

Fig. 3.2 Basic science-technology cycle for macro-innovations
Source: Meyer-Krahmer and Dreher, 2004.

of the resulting alienation, only those actors with the greatest endurance or radical new approaches contribute to the reorientation of the technology's development (stage 4). They achieve critical industrial breakthroughs. The breakthrough which is the fastest to achieve market acceptance shapes the future handling of the technology (dominant designs) (stage 5). In the phase of diffusion (stage 6), applications expand again because economies of scale result in price reduction and allow new application areas and low-cost markets to be tapped.

In each of these phases, it is possible to identify typical actors, different companies as well as technology management and policy strategies (e.g. by patent data analysis within the first four or five stages of the technology cycle, by citation analysis) and to identify other typical actors for the rise and diffusion phase by market penetration data.

3.4.2 Empirical Analysis and Indicators

An empirical approach is used to describe and assess technology cycles based on patent and publication statistics. The analysis of patent applications has proven a valid tool for describing the development of technologies.[6] Patents reflect the inventive activities in the technology fields considered, and a substantial share of about

[6]For further details, see Hinze and Schmoch (2004).

40–60% of the patents are actually used for innovations, that is for new products or processes in the market (Grupp, 1998: 183). So in standard technology-push models, the patent activity should be visible before the introduction of the corresponding product or process into the market.

A specific advantage of patent statistics is the possibility of tracing technology activities for long periods on the basis of patent classification codes. In the case of US patents, this can be done using current classification codes since all documents are retroactively reclassified as in electronic databases. A major disadvantage of patents at the US Patent and Trademark Office (USPTO), however, is the relative dominance of US inventors, so that technology trends primarily show US-based activities. This so-called domestic advantage and the associated national bias are characteristic for all national patent offices. An alternative is to use applications at the European Patent Office (EPO) as the distribution of countries is much more balanced here. At the EPO, there is a certain bias in favour of European countries, but the number of US and Japanese applications is still considerable, and the United States are still the leading country (see Hinze & Schmoch, 2004). To summarise, it is largely correct to assume that patent trends at the EPO reflect trends of technology-oriented innovation activity on a global level.

At the EPO, technology fields can be defined using the codes of the International Patent Classification (IPC); however, the IPC is not retroactively reclassified like the US Patent Office Classification (USPOC). But internally, the EPO examiners use the European Classification (ECLA) which is more disaggregated than the IPC and is retroactively reclassified. The ECLA codes are publicly available in the database PLUSPAT (Orbit-Questel) and make it possible to do long-term analyses at the EPO as well as at many other offices. A specific restriction of EPO analyses is that this office was only established in 1978 and until the end of the 1980s we can observe a steady transition from direct foreign applications at national European offices to central applications at the EPO. In the present situation, the vast majority of foreign patents are filed through the EPO, and the number of direct foreign applications at domestic offices is negligible. In consequence, it is possible to observe internationally balanced technology trends using EPO applications from about 1990. In the period between 1978 and 1990, foreign applications have to be recorded at the EPO and the national offices in parallel, and before 1978, only national European patent offices are available with the aforementioned distortions due to the respective domestic advantage.

The relevant scientific activities were mapped by publication statistics in the online version of the Science Citation Index (SCI, host STN). The SCI covers worldwide publications from about 1975. The basic problem of the publication searches in this analysis was to identify scientific activities which appropriately reflect the contents of the technology fields considered, so the concordance problem had to be solved (Grupp, 1998: 159ff). As the classification of the SCI is rather coarse, the searches were primarily based on combinations of keywords. With regard to specific areas, e.g. conductive polymers, keyword searches proved to be less appropriate in the SCI, and the publication analysis was done using the database PASCAL (Questel-Orbit) which has a much finer classification. The standard reference years for publication analyses are publication years. However, in the patent analyses, the authors used the priority years, i.e. the year in which the first application was made, in order to be as close as possible to the time of technology generation. Then the time series for publications were deferred by one year in order to take into account the time lag between the submission of a paper at a journal and its publication, due to the referee process.[7]

Finding appropriate economic data is even more difficult, as economic product or sector classifications often cannot be linked to technology fields due to the differing underlying classification logic. Economic data are not retroactively reclassified after changes in classifications, and in particular, economic classifications do not reflect recent trends, so that new fields cannot be displayed. Several indicators can be used such as diffusion data, market analysis of customers and customer potentials, forecasts etc. All in all, the introduction of economic proxies is problematic and not totally satisfying, but still feasible.

3.4.3 Lessons from Previous Assessments

In a technologically broader study, Dreher et al. (2005) analysed technologies such as robotics, manufacturing lasers, immobilised enzymes and conductive polymers. The examples followed the double-boom character of the technology cycle and displayed similar patterns which may be described as follows:

- All fields are complex, science-based fields, and technical activities (patents) are closely linked to scientific ones (publications). Scientific activities extend throughout the entire observation period and even increase over time.
- Scientific activities start at the same time as technical ones or a few years earlier. They can be qualified as oriented towards either basic or applied research.
- The results of pure, basic, curiosity-driven research in the fields analysed often appear several decades before the first patent boom, but are not taken up by either technology or science. The fundamental problem is obviously one of realising the basic theoretical concepts even on a laboratory scale. If the stage is reached where the transfer to technology seems realistic, then the relevant scientific and technical activities begin immediately.
- In the course of the technology cycles, scientific trends always precede technical ones by five to ten years. They enter a stagnation period even before the first patent boom and pick up again distinctly before the second patent boom.
- Scientific activities are more stable and do not display the pronounced undulations of the technological ones. Firms react quickly if the expected technological and commercial results are not achieved within a short period.

---

[7] So it was assumed that the articles were written about one year before their publication in order to achieve a reference equivalent to the priority year of patents.

- The technology cycle is quite long. The lag between the first and the second increase may be 15 years or more. In all the investigated cases, the peak of the second boom is not yet visible, so that a total length of thirty or even forty years may be realistic.
- The market activities reach a substantial level, but not until the beginning of the second technical boom.

The first boom can be labelled the "science-push" or "technology-push" cycle, as the detection of new basic concepts triggers substantial technological activity. As the technological and scientific activities are undertaken almost simultaneously, it is not possible to make a clear distinction between science-push and technology-push.

Since the fields are complex, they often run into problems which cannot be solved in the short term. From the perspective of firms, the period of five to eight years between the start and the peak of the first boom is obviously quite long, and they may react by decreasing their research or even dropping out completely. In this context, a further observation is that only a few firms appear as patent applicants in the first as well as the second boom. Against the background of engagement in a risky area, actors try to imitate other actors, in particular those with good reputations. As a result, if leading firms decide to enter a new technology field, others will soon follow; if, later on, leading firms exit this field, others will leave as well.

The first boom can be linked to the hype cycle of Gartner or the first stage of Rickerby and Matthews. The period of increase can be described by a negative feedback loop: unsatisfactory research results are an incentive for increasing research activities. After the peak, the feedback becomes a positive characteristic. Unsatisfactory technological (and market) results trigger a drop in research. So the characterisation of the first boom as a science/technology-push cycle does not imply a simple linear sequence.

The second boom is initiated by solutions being found to the crucial problems which led to the end of the first boom. Obviously, these solutions are achieved in science, an event, which is immediately recognised by other scientists. In contrast, the firms seem to be more sceptical after their less favourable experiences of the first boom, and prefer to wait until the new solutions prove reliable. The thesis sometimes put forward – that firms are not aware of this breakthrough – is less probable.

In a more general perspective, the first boom may be associated with a high level of uncertainty and a parallel engagement of scientific and industrial laboratories for elaborating basic concepts and solutions; the second boom is characterized by a higher level of certainty and a strong focus on market demand. With regard to the related actor structures, there are striking parallels to Callon's (2001) concept of emergent and consolidated socio-economic networks. The major difference, however, is that the scientific activities are still substantial during the second consolidated phase, albeit with new targets. In general, the findings support the initial general model developed by Meyer-Krahmer and Dreher (2004).

3.4.4 Policy Implications

The quite long cycles involved imply that supporting a field may have different impacts depending on the respective stage of development. Policy administrators often lean towards supporting new areas in the early stage of euphoria and feel themselves deceived if visible results are not achieved within five years. However, new fields should be supported particularly during the stage of disillusionment when industrial firms tend to slow down their activities. This is why public support should be designed for the long term. In any case, the drop in patent activities is not a definite indication that further support is superfluous. On the contrary, innovation policy should look more closely at decreasing patent activities in order to be ready to back the start of the second boom relevant for market growth. In this context, it should be helpful to observe the related scientific activities in order to reduce the risk of lost investments.

The double-boom characteristic of complex fields should be kept in mind when planning the support of promising fields. For instance, it is obvious that large parts of nanotechnology are still at the stage of basic research, thus at the stage of understanding basic characteristics at the nano level, and the share of applicable results is still limited. So it would not be surprising if the present euphoria is actually linked with the first cycle of a double-boom characteristic and that a period of disillusionment will follow in a few years. Well-designed public support may shorten the intermediate phase of decrease.

Firms should be careful about getting too involved during the early stages and should try to avoid the following unrealistic euphoria. But they should recognise breakthroughs and be prepared to take the risk of investing before the second boom in order to benefit from first mover advantages. The relevant periods for application-oriented research institutions are the stages of disillusionment and reorientation before the second boom.

It is less clear at which stage research institutions with a more basic orientation should get involved. However, basic scientific (and technological) activities are still vital to opening up new applications and markets as well as new areas of scientific research.

Of course, to achieve short-term market success, it is best to invest at the beginning of the second boom. At this stage, public support might be useful to prepare new application areas. Nevertheless, public authorities should not neglect the support of riskier fields at earlier stages because industry is reluctant to get involved here at this time. Furthermore, the undulating development of technological activities has consequences for long-term technology predictions. It is obviously not sufficient to merely look at recent technology growth. Instead, it is necessary to consider longer periods and to link growth analysis to specific cycle stages. In addition, technology paths have to be linked to the corresponding science paths. In order to make more specific recommendations on a policy level, more information on interaction patterns, actor's behaviour etc. is necessary. This is one of the conclusions resulting from analyses of the innovation system and should be combined with the findings on the stage of development to produce recommendations for the timing and targets of public support.

Any interpretation of the indicator data should be checked by technology and market experts in the field. Especially if quantitative data on patents, publications, cost reduction potentials, or market shares are not easily available, to a certain degree it is still possible to gain an insight into the status of the technology cycle by interviewing specialists. This option is, of course, less transparent and carries the risk of reflecting the opinions of the interviewees rather than the objective status of the technology considered. However, it may be useful to provide initial information which can be used to limit the efforts necessary if an analysis of competing technologies has to be made. Furthermore, different kinds of experts must be consulted in order to identify the bias of their contextual situations.

3.5 Innovation System Approach

This section introduces the underlying concepts of the second methodological pillar on which the EduaR&D process rests: the innovation system approach.

3.5.1 Definition of Technological Innovation Systems

The starting point for the concept of an innovation system is its functional definition: innovation systems generate, diffuse and use innovations (e.g. Carlsson, 2002, p. 235). The conceptual description of the innovation system approach employed in the EduaR&D process draws largely on Carlsson and collaborators and on their discussion of technological innovation systems. Innovation systems are comprised of components[8] and relationships, where components are the working parts of an innovation system and relationships are the linkages between the components.

Components of an innovation system consist of (1) actors, such as individuals, firms, organizations, institutions and groups (collections) of individual actors such as associations etc., (2) tangible and intangible artefacts such as technologies, modules or products, (3) institutional regulations and framework conditions.

Relationships between the components can be contractual, hierarchical arrangements in companies or administrations, informal dependencies, communication, intended or unintended technology or knowledge transfer, direct relationships such as direct contacts between actors, or indirect relationships such as competitive relationships moderated by the market. Both components and relationships can be characterized by their properties and attributes.

---

[8]Where the meaning is unambiguous, we use the term *component* instead of *innovation system component*. Where ambiguity might occur we use *technological component* to denote a part of the technology under analysis and *innovation system component* to denote a part of the innovation system.

The dynamics of and within the innovation system are triggered by changes in the set of components and relationships within the system – new components appear and existing components disappear over time; components enter new relationships or relinquish old ones. System dynamics can also be caused by changes in the attributes of both components and relationships. The literature distinguishes among different innovation systems:

- *National innovation systems* describe the set of all components and relationships relevant for the development, diffusion and application of new knowledge and technologies in a specific country. Hence, the system is delineated by national boundaries and neglects sectoral and technology-specific aspects.
- *Sectoral innovation systems* are based on the economic demarcation of sectors. In most cases it takes the sectors as given, although knowledge and technological solutions tend to permeate through sectoral boundaries.
- *Technological innovation systems* describe the set of all components and relationships relevant for the development, diffusion and application of a specific technology. This focus coincides with the goal of the EduaR&D process to analyse the technology-specific system for individual new energy technologies.

Technological innovation systems (TIS) are usually not restricted by national or sectoral boundaries. If the development dynamics of the analysed technologies unfold on a multinational or global scale and involve several sectors, the description of the TIS covers global and multi-sectoral actors. To arrive at a thorough understanding of a technological innovation system, one has to analyse four dimensions (see: Carlsson et al. 2002, p. 10):

- the cognitive technological dimension,
- the economic dimension,
- the organizational and institutional dimension and
- the societal dimension.

3.5.2 Key Questions and Methodological Approach

This section elaborates on the above dimensions to direct the innovation systems approach within the EduaR&D process.

3.5.2.1 Cognitive Technological Dimension

The cognitive technological dimension covers the generation and dissemination of knowledge required for the development, diffusion and application of the technology. Key questions to be covered by the analysis include (see Exhibit 3.3):

- How do the technology and its associated products, modules and technological components work? Which part of the technology poses problems and challenges to the development, application and diffusion of the technology? Are the products associated with the technology marketable?
- What are the essential items of knowledge required for the technology? Where do the relevant scientific insights originate? Where does the inspiration for new developments come from? Which disciplines have to collaborate to generate new knowledge for the technology? Which mechanisms are working in the innovation system to select ideas and approaches? How does new knowledge diffuse within the system?

Exhibit 3.3 Approach: Cognitive technological dimension

To analyse the functions and the technical challenges of the technology applications examined and their knowledge stock, it is necessary to conduct interviews with specialists (or for EduaR&D to refer to the in-house specialist knowledge of the FhG-ISI) and perform literature analyses. These kinds of discussions and literature analyses have to be done early on in the analysis in order to obtain a picture of the technological drivers (opportunities and actors as well as obstacles) in a technology innovation system. Besides the methods used here, cited reference searches in patents and publications can also be conducted to trace the diffusion of the necessary knowledge and the links to the respective knowledge stocks and knowledge holders.

3.5.2.2 The Economic Dimension

The economic dimension essentially focuses on the economic utilisation of knowledge and technologies. The definition of the market not only relates to the final product but also encompasses the modules and technological components required for the technology and the relevant actors.

The economic dimension can be characterized by the cost of the technology (also relative to competing technologies), the market structure and the conduct and performance of actors (see Exhibit 3.4). Depending on the respective technology cycle phase, various indicators can be used to characterize the economic dimension. The operative possibilities of the actors are largely influenced by the financing conditions for the technology such as the availability of research grants, public funding for R&D, or venture capital. Beyond the conditions shaping the creation of the technology, the manner in which the technological innovation system functions is influenced by the actual or potential (final) demand for the products, modules and components related to the technology. A host of microeconomic indicators can be used to characterise the demand and its development.

Exhibit 3.4 Approach: Economic dimension

To define and analyse the economic dimension it is necessary to evaluate cost analyses (or cost projections) of the technical solution regarded, cost comparisons if technology substitution is involved, existing sector analyses and company analyses and to hold discussions of these issues with leading actors. The use of energy system models and cost reduction potential analyses, which are applied more extensively in other research projects running parallel to EduaR&D, are valuable analysis instruments for cost estimation, the foreseeable economic competition and possible demand.

Generally the economic conditions within the technological innovation system are not independent of the distribution of knowledge since knowledge represents a key asset in the development and diffusion of technologies. In addition to a static characterization of the status quo, integration of potential future aspects of the economic dimension is also required for an understanding of the technological innovation system.

3.5.2.3 The Organizational and Institutional Dimension

This dimension covers the relevant actors and the institutional arrangements within the technological innovation system (see Fig. 3.3):

- *Public research* comprises actors such as universities, public research organizations and publicly-funded research organizations. Generally these organizations

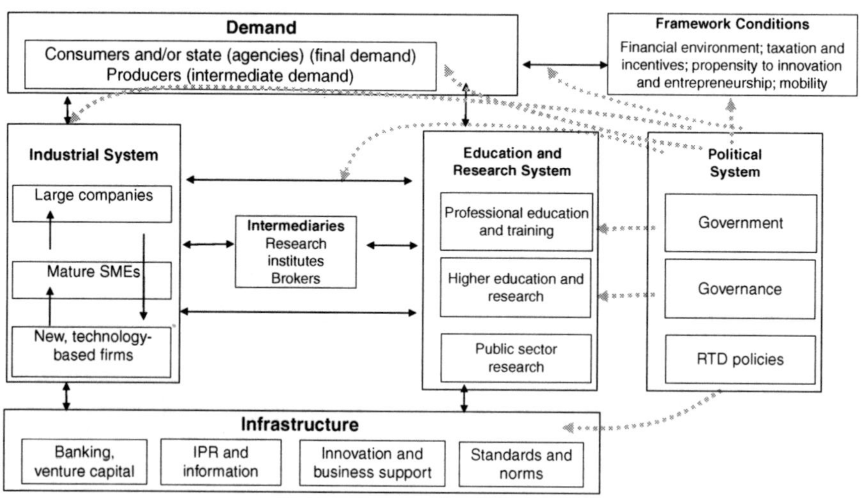

Fig. 3.3 Outline of an innovation system of a technology
Source: Technopolis 2000, modified and extended by S. Kuhlmann/J.Edler, ISI.

can be located on the continuum between basic research[9] and applied research.[10] The more knowledge-intensive a new technology, the more favourable a close relationship between actors with a basic research orientation and actors with a more applied focus.

- *Industry* develops, produces and finally supplies the technology and related services to the market. Industry laboratories, workshops and engineering offices are the key players prior to the commercialization of new technologies. It seems that the more complex a technology is, the greater the need for collaboration between industrial actors.
- *Intermediaries* are a rather heterogeneous group of actors. They consist of technology transfer organizations, standardization organizations, sources of corporate finance such as venture capitalists and technology media. As a rule, these institutions are not directly involved in the development of new technologies but they do support their development and commercialization.
- Industry, households and procurement by public agencies provide the *demand* for the new technology. Their capabilities and their willingness to pay for the new technology are key determinants for its diffusion and commercial success.
- *Governmental actors* supply funding to other actors in the innovation system, regulate the innovation environment, and influence the future demand for the new technology. Beyond this, governmental actors have access to a number of different measures used to set incentives for new technologies. The EduaR&D process offers these groups of actors strategic intelligence since it screens and analyses the innovation system, the properties of the technology (also a component of the innovation system) and the future development and relevance of the technology.[11]
- Other relevant players may be actors or actor groups who influence the acceptance or the regulatory environment of new technologies by influencing the awareness of certain benefits or risks associated with the technology.

For all actor groups there are also associations such as industrial associations, consumer associations etc. which aggregate individual attitudes and interests and work as forums to develop a "common understanding" of the new technologies.

---

[9] In Germany, e.g. universities, the Max-Planck Society.

[10] In Germany, e.g. the Fraunhofer-Society, the Helmholtz-Society, AiF-Research Organizations.

[11] In case the reader is more comfortable with a single approach rather than three different ones, the innovation systems analysis can be seen as the overarching approach under which the energy-economic analysis and the technology life cycle analysis can be subsumed.

As the technology itself is a tangible (or intangible) artefact, it constitutes a key component in the technological innovation system. The energy-economic analysis investigates current and future properties of the technology as one component of the innovation system. The technology life cycle also – from a different angle and with a different conceptual background – analyses the properties of the technology as a component of the innovation system. Both the life cycle analysis and the energy-economic analysis can be seen as part of the technology innovation system analysis.

In addition to the actors, the organizational-institutional dimension also includes institutions such as the regulations, norms, habits and traditions shaping the behaviour of and the relationships between actors (see Exhibit 3.5).

Exhibit 3.5 Approach: Organizational institutional dimension

Three approaches have to be combined to methodically define and analyse the system of actors:

1. *Patent and, where necessary, publication analyses to determine the main actors in the field (assumption: high output activities signal a certain significance) and to identify cooperation intensities (co-inventors, co-publications).*
2. *Context analysis based on the main documents of relevant institutions (including sponsors), secondary literature, popular and scientific journals etc. As long as the assignment is unambiguous, this can be based on analyses and descriptions of the sectors in which the technology regarded is being or is to be applied.*
3. *Interview programme: definition of specific questions for each actor group. In-depth interviews with at least one representative from each actor group (there is an example of a list of questions for the manufacturers of SOFC fuel cells in the Annex). Criteria for the selection of interview partners are: relevance, a certain degree of representativeness for their actor group as well as a broad overview of the field.*

Not only the competences, capabilities and features of each individual actor shape the development of technologies and products but also the links and relationships among them which jointly create the emerging ability of the system to solve problems which cannot be solved by individuals acting in isolation. To achieve this characteristic, the countervailing forces within the system have to be reduced.

The key questions for the analysis of a technological innovation system are:

- Identification of the relevant actors. Who drives and who slows down the development process?
- Identification of the relationships crucial for the generation, dissemination and application of knowledge.
- Identification of the relationships which determine the time to market from a new idea to the market launch of a new product.
- Identification of the most efficient relationships of governmental and public actors to reduce the developmental obstacles.
- Identification of the properties of the relationships
    - formal – informal,
    - persistent – ad hoc,
    - transparent – non-transparent,
    - market driven – not market driven;
    - cooperative – competitive

3.5.2.4 The Societal Dimension

The societal dimension does not focus on societal targets as such but investigates to what extent the technology contributes to achieving given societal goals. The underlying assumption is that these societal goals exist independent of the analysed technology. The societal dimension also covers how social phenomena influence the development, diffusion or acceptance of new technologies (see Exhibit 3.6).

Exhibit 3.6 Approach: Societal dimension

To a large degree, the analysis of the societal dimension is covered in the energy-economic analysis discussed above. The contribution to goals other than energy-economic ones such as employment, growth and regional-economic goals can be derived from the analysis of the economic dimension or is part of the assessment of the ancillary effects of this dimension.

3.6 Integration of the Methodologies and Policy Aspects

The value added of the methodological approach developed in the EduaR&D process is the integration of the three conceptual elements and their mutual synthesis of results and observations. This is shortly illustrated in Sect. 3.6.1 by answering policy questions by the three types of methods integrated here in the EduaR&D process.

3.6.1 The Contributions of the Methodological Approaches to Policy Design

The integration of the energy-economic analysis, the technology life cycle analysis and the innovation system analysis in the EduaR&D process (see Table 3.9) is performed in an iterative way where the results of each concept may influence the search and evaluation process of the other two. This interplay of results between the three approaches seems to be (and is) rather labour intensive. However, more comprehensive knowledge about the new technology, its energy economic importance, its position on the technology cycle and its innovation system may be quite useful for policy decisions and for steering the R&D process in a target-oriented manner though critical phases of the R&D and innovation process.

3.6.2 Market and System Failures and Policy Instruments in the Technology Cycle

A major objective of EduaR&D is to define concrete policy instruments for the effective support of a selected technology. To select and tailor appropriate

Table 3.9 Combination of the three approaches in the EduaR&D process

Policy questions	Energy-economic analysis	Diffusion approach	Innovation system approach
What?	Determination of the energy-economic significance of the new technology/ies, the necessary technological progress and cost level which could or should be achieved. Indications of diffusion barriers.	Determination of the position in the technology cycle (patent analysis), definition of the technology at a "suitable" point, publication analyses (dynamics of the relevant knowledge production), market potentials.	Determine which concrete level of the technology (module/product/system) and which system elements of this technology should be sponsored.
Who?	Identify the energy users affected and technology attributes, the possible R&D-funding institutions and possible research institutions.	Assessment of the main actors as a result of the patent (and possibly publication) analysis, assess the type of actors (users, consumers) which have to be approached.	Identify the main actors in the respective subsystem and their needs as well as possibilities of sponsorship leverage.
How?	First indications of the prospective instruments for R&D or market diffusion (only tentative prior to the diffusion analysis and the innovation system approach).	The distribution of key knowledge holders and technology drivers may reveal possible gaps in cooperation. Market analyses make it possible to assess the ability of the market to take up knowledge.	Definition of the processes and activities which could/should be induced and reinforced via the sponsorship, i.e. via the conditioning of sponsorship instruments. Comprehensive approach results in large margins which may not fall under the portfolio of a specific ministry.
Methods	Use of energy system analyses and projections, analyses of cost degression potentials.	Patent and publication research, measured values for installations, units sold, projections of market diffusion.	Modelling the subsystem, definition of the most important elements and then also concrete actors. Definition of actor- and technology-specific questions, interviews.

instruments, this chapter provides a selection of policy instruments derived from a combination of the innovation systems approach and the technology cycle. It defines the functionalities of technological innovation systems and the potential system failures – or bottlenecks – for each of the six stylised phases of the technology cycle and describes policy measures that are suitable for supporting the innovation system's functionality or for patching a system's failure within a specific policy phase. Thus, it is not only the phase or the failure that defines the need for a policy measure, but the combination of phase and system failure.

We have applied a broad understanding of "failure", including the traditional, neo-classical *market failure* concept. In the traditional, neo-classical literature, the legitimation for public action in research was based on a market failure. This well known concept relates to the problem that investment in *basic* research will be too low due to an inadequate appropriation of benefits emanating from unavoidable spillovers to other market actors and a high risk of commercialisation and thus a limited return on investment. Public budgets have to step in and provide or finance basic research (Arrow, 1962). If they do not, not only would there be too little basic research, but this would be concentrated only on selected, most promising technological areas. Thus, in the EduaR&D process, it is sufficient to refer to the necessity of giving basic researchers in the system institutional funding or appropriate project funding in order to enable constant knowledge production as one source of future innovations.

Possible *system failures* are derived from the concept of innovation systems explained above. The basic argument is that for the generation and diffusion of innovations, a certain set of functions is required that has to be fulfilled by innovation systems.[12] A technological innovation system must deliver the knowledge required for the understanding and development of a certain technology that is able to fulfil market demands. This comprises *basic* and *applied research*, as well as appropriate *education* and *formation* (cognitive dimension). Furthermore, alongside the agents generating basic and applied knowledge, there must be both the *economic interest* and the *capabilities* to *co-develop*, *absorb*, *process*, *apply* and *exploit* this knowledge in the market.

This entails not only the *development* and *production* of new technologies, but also the associated service and maintenance. Furthermore, the market needs to be ready, with regard to both the *willingness and absorptive capacities of users*, and the *infrastructure, complementary technologies and services* needed for a broad application (economic dimension). Finally, the various elements of the system need to be functionally connected, i.e. appropriate *networks* are needed in all phases and for all functions of the system (interaction dimension). These interactions may vary widely with regard to their form and the actors involved in the different phases, but there is a strong need for interaction in each phase.

---

[12] For an elaboration of system functions, see Hekkert, Suurs, Kuhlmann, and Smits, (2006): Functions of Innovation Systems: A new approach for analysing technological change; in: Technological Forecasting and Social Change (forthcoming).

A more detailed overview of the functions of a technological innovation system reflects the need to produce knowledge, technologies and innovations and diffuse them into the market (see Table 3.10). The functions, failures and instruments are structured according to the six phases of the technology cycle. However, it is obvious that the attribution of needs, failures and instruments to the phases is undertaken according to the phase in which this need, failure and instrument is most obvious and relevant. It does *not* imply that the application of the instruments should be strictly limited to a certain phase.

For example, not all innovations adhere rigidly to the six phases given. Especially with modular innovations, which include relatively few novel technologies but tend to be developed based on an innovative combination of known elements, there may be a mixture of different phases. For example, while CO_2-free power stations are generally still in phase 2 of the technology cycle, actors are already taking market aspects into account which would only be expected in phases 4 and 5 because they can foresee relatively clearly how the technology will look in the end and what its market will be. In such cases, political actors should check the applicability of measures of later phases and may also conclude on particular research issues such as acceptability or acceptance of a technology still in the R&D process.

Similarly, policies that have priority in a certain phase may still continue to be applied even if the technology has reached another stage of the cycle. For example, sometimes even in the "rise" phase, in which application and demand orientation is most important, it is still relevant to fund basic research in order to overcome persistent technological problems, as well as the associated necessary co-operation schemes. As a rule of thumb, the instruments of its neighbouring phases will always be important for a specific phase, so there needs to be an instrumental overlap between phases. This overlap is not demonstrated in the table, which has to focus on the *major schemes in each phase*.

Moreover, (see Table 3.10) no distinction is made according to technology, sector or market condition. There is a general attribution of instruments to phases that will require adaptation depending on the technology and context concerned. For instance, software technologies may be produced by many small-scale companies in parallel, while technologies requiring high investments may be limited to large, multinational companies. These variants require very different policies, especially in the 3rd–5th phases. By the same token, the policy matrix cannot differentiate all the potential target groups in detail, and the final selection and definition of policies based on this matrix will have to take into account the specific needs of certain target groups such as for example small and medium-sized firms, and adapt the policies accordingly.

This heuristic should help to detect system failures and to define appropriate policy instruments. However, the guidelines may not be able to cover all potential phase-failure combinations. Thus, while policy can learn from this method, it should remain flexible whenever certain failures occur that cannot be clearly attributed to a certain phase.

We also need to stress that while this policy tool box aims to help decision makers to think systematically about policy choices, it could and should never be applied

Table 3.10 System function, system failure and policy instruments in the technology cycle

System functions and pre-conditions	Potential system failure	Policy instrument to address this failure
Phase 1 Invention	• Lack of R&D funding and other financial incentive schemes (e.g. public research funding; institutional funding, indirect support via tax deductions etc.) • Rigid programmes with a bias towards established routines and approaches	• **Institutional funding** of basic public research in universities and non-university institutes • **Science support programmes** that encourage **risky, highly innovative research** to circumvent established scientific paths (built-in failure tolerance)
Generation of knew, basic knowledge as the basis for new applications and solutions	• Insufficient infrastructure or unfavourable organized admission to infrastructure (e.g. large scale research infrastructure) • Unfavourable regulation (e.g. with regard to stem cells) • Unfavourable culture within the basic science subsystem (e.g. total dependence on leading professors, "Professorenhörigkeit")	• Funding of necessary **infrastructure** • Check and improve **regulations**, moderate public discourse to enable regulatory adaptations • Implement institutional changes in the science system to **induce scientific variety**
Production of talent to secure long term production of knowledge	• Too few national basic researchers in certain fields, lack of skills and/or methodology in an area indispensable to a specific technology • Too few scientific talents, inadequate education • Low attractiveness in the international arena	• **Attraction** schemes, dismantling of bureaucratic and institutional hurdles for science immigration • **International mobility schemes** • **Attractive tenure** tracks for young scientists

Table 3.10 (continued)

System functions and pre-conditions	Potential system failure	Policy instrument to address this failure
Composition of formerly separated knowledge packages (on national and international scale) Joint production of knowledge, interdisciplinary and as co-operation between basic and applied public researchers (mode 2 of knowledge production)	• Departmentalisation of basic research, lack of trans-disciplinary or trans-institutional research activities • Low (cultural, financial) incentives for international cooperation	• Funding of **interdisciplinary research groups** to enable composition of knowledge and methodological approaches and mutual fertilisation • Provisions to enable **international collaboration** of basic researchers (within DFG funding and BMBF and BMWi programmes) in order to ensure absorption of globally produced knowledge and benchmarking of national researchers • Centralization of research funding for each technology at one governmental institution (e.g. in one ministry) or **strong co-ordination** between ministries
Phase 2: Euphoria High level of basic research but with a focus on closer interaction between basic and applied research	• Lack of application or "problem" perspective in "basic" research • Strong fragmentation of parts of the system and lack of interaction	• Dedicated **top-down programmes** in basic research • Schemes to enable **more intensive co-operation between basic research and application-oriented research**, i.e. improving the conditions for non-university, non-Max Planck Society (MPG) researchers to collaborate with basic researchers, e.g. technology parks

Table 3.10 (continued)

System functions and pre-conditions	Potential system failure	Policy instrument to address this failure
Streamlining research activities into areas relevant for industry – and making industry adjust to new possibilities offered by new technologies (mode 2 of knowledge production, application end)	• Lack of absorptive capacities in industry to understand the progress in basic research, to contribute to basic and applied research, to recognise innovation potential and – above all – to modify new technologies into marketable products • Lack of knowledge in industry about public research activities, lack of monitoring • Unfavourable IPR • Unwillingness or inability of big industry to depart from business-as-usual	• **Dedicated research labs** co-funded by industry and state[13] • Support of **structural co-operation** between science and industry such as competence centres or temporary labs and institutes[14] • **Programmes for co-operative research** with a high focus on scientific excellence, but **with a view to industry's interests** • **Mobility schemes between industry and science** • Support of industry-defined basic research topics[15]

[13]Examples of such labs could be the Austrian Christian-Doppler-Labs (Schibany et al. 2005) or the US-University-Industry Collaborative Research Centres, both of which are supported by a group of companies jointly interested in a certain basic research line. The funding includes setting up excellent post-doc research teams attached to the core university activities.

[14]In the competence centre approach, a set of university institutes and companies join forces and de facto establish joint institutes, oriented towards certain technologies, mainly inter-disciplinary and inter-sectoral. These centres define research programmes, organise horizontal, basic research projects for the whole institute as well as small scale, bi- and multi-lateral projects. The main feature is the mid-term co-operation that combines the strategic and operative action of a group of companies and institutes. For a description of competence centre programmes, see: Good practices for the management of Multi Actors and Multi Measures Programmes (MAPs) in RTDI policy; http://www.map-network.net/publications/the-handbook.pdf.

[15]As in the German AiF scheme in which so-called Sektorale Industrielle Forschungsvereinigungen (Sectoral Industrial Research Associations) define topics for basic research or applied basic research issues (modified AIF).

Table 3.10 (continued)

System functions and pre-conditions	Potential system failure	Policy instrument to address this failure
		• Continuously checking and improving IPR regulation • Fostering close interaction between political, industrial and research actors to combine efforts
Match research areas and future market needs to direct euphoria into diffusion as quickly as possible	• Lack of knowledge and market intelligence on the absorptive conditions in the early lead markets • Research may not be focused optimally	• **Constructive technological assessment**[17] and foresight activities in order to ascertain very early on the probability of complementary technologies (supply foresight) and development of demand (demand foresight) • **Monitoring international activities** in order to detect potential foreign lead market possibilities • Identifying bottlenecks and focusing research funding on them
Transfer results of blue sky research into the application arena	• Lack of transfer institutions, counterproductive incentive structures in public research • Lack of vertical networking	• **Support of patenting and licensing** from basic research institutions, support of intermediaries • **Support of networks between knowledge providers and actors exploiting knowledge** to streamline activities along value chains

[16]Constructive Technology Assessment is a discursive, moderated activity whereby the generation and design of a technology is influenced and shaped by societal needs, expressed through key stakeholders and representatives of society. CTA thus complements the supply (i.e. science and/or industry driven) definition of technologies. Sundermann (1999): Constructive Technology Assessment, in: Stephan Bröchler, Georg Simonis, Karsten Sundermann (Hg.): Handbuch Technikfolgenabschätzung, Berlin 1999, edition sigma, S. 119–128.

Table 3.10 (continued)

System functions and pre-conditions	Potential system failure	Policy instrument to address this failure
Phase 3: Disillusionment		
Persistence of research activities (staying power)	• Lack of direction and patience to define and overcome bottlenecks • Uncertainty about future economic, political, and regulatory foundations	• **Review technological and market bottlenecks**, including review of competing technologies and their contexts • **Focused, but persistent research programmes** • **Analyse regulation required** (and other boundary conditions) and implementation, if possible
Science-Industry interaction with a view to market possibilities *and* bottlenecks	• Interaction too broad and non-directed, underestimating technological and market risks	• **Cooperative and collective (AiF) research programmes** • **Indirect-specific promotion** of applied development activities on **supply side** • Support co-operation between development oriented companies and institutes to streamline activities
Selection of sustainable technology paths	• Too many technological variants are developed, dominant designs need to be defined	• Support **pre-standardization activities** such as measurement technologies, references and performance parameter definitions but no product standard at this time • Support activities to filter promising technology paths, e.g. by developing roadmaps and/or scenarios together with different actors
Fine tuning, adapting and testing technologies	• Technologies are prematurely induced into the market or functionalities are not demonstrated to potential users	• Support **testing and demonstration facilities**

Table 3.10 (continued)

System functions and pre-conditions	Potential system failure	Policy instrument to address this failure
Financial support for daring start-ups and young companies throughout the turbulence of disillusionment	• Venture capital is scarce in times of disillusionment, but crucial for those scientists and companies with a potential to score in the market at a later time	• **Pre-seed and seed financing** with a very strong check of market and technological bottlenecks
Phase 4: Re-orientation		
Strategic intelligence as to the niches of a new technology left after phase 3 as well as regulatory provisions conducive to the market introduction of technologies, reconciling needs of society, processing of risks and political considerations with technological possibilities	• Actors lack confidence in and knowledge of certain market possibilities	• Lead **market studies**, studies on competing technologies • **Focused technology assessment** • **User involvement** for the selection and definition process of the technologies (CTA) • **Scientific councils** to advise governments • **Regulatory foresight**, ex ante regulatory impact assessment • Establishment of **IPR** (more in phase 2, Euphoria) • Support of **standardisation processes**
Venture capital needed for early market introduction	• Banks and other venture capital donors are extremely cautious	• **Venture capital** via public credits to close the financing gaps after phase 3, but credits need to be based on clear market analysis and business plans
Cooperation of different disciplines needed to actually build innovative applications	• Different cultures of different disciplines hinder efficient communication and cooperation	• Foster interdisciplinarity as early in the process as possible, e.g. by specially designed research funding

Table 3.10 (continued)

System functions and pre-conditions	Potential system failure	Policy instrument to address this failure
Focused research, market oriented systems integration	• The system creating and transferring knowledge struggles with focusing its research activities on the promising technological path	• Support **co-operative research** concentrated on **solution of clearly defined bottlenecks** and connected to well proven market potential • Support of **co-operation between producers and users**
Awareness on the supply and the demand sides about what finally works in technological terms (and thus in terms of costs)	• Scepticism throughout the technological system, need for conviction	• Support activities **demonstrating** that the technology works reliably (e.g. demonstration plants) • **Awareness measures**
Political decisions as to whether a technology needs to diffuse for social, environmental or any other political reason; if this is the case, corresponding measures need to be designed to pull technologies in the market	• Unclear signals from the political arena about the willingness to support a new technology for political reasons	• **Early public procurement measures** • **Bundling of innovative demand** (co-operative procurement) • **Financial support** for private demand • **Regulation to prescribe market structures** (e.g. feed-in tariffs (EEG), carbon dioxide emission certificates)

Table 3.10 (continued)

System functions and pre-conditions	Potential system failure	Policy instrument to address this failure
Phase 5: Rise		
Broad market acceptance crucial for take off, with a positive long term cost-benefit ratio	• Demand may not take off for a number of reasons such as: high entry costs of technologies, high switching and transaction costs (lock-in effects), lack of general awareness and trust, lack of network effects in early phases • High search costs • Economic risk for amortization of investment in the technology. • Low public acceptance of an innovation might hinder its promotion	• **Public procurement and bundling of demand** (co-operative procurement) • **Broad awareness measures** including labelling, information and support for marketing campaigns – depending on the political will to push a certain technology and public concerns • Establish or facilitate information and collaboration networks • **Development of new (potentially PPP) models for operating** the technology (esp. utilize the cost savings generated to pay back the additional investment costs), development of new financial instruments (to balance the economic risk by hedging energy prices)

[17] For a broader discussion of system failures and demand see Edler (2006): Public action to spur innovation via demand (working title), in: Smits, P/Kuhlmann, S./Shpaira P. (eds.), The Co-Evolution of Innovation Policy - Innovation Policy Dynamics, Systems, and Governance (forthcoming).

Table 3.10 (continued)

System functions and pre-conditions	Potential system failure	Policy instrument to address this failure
Continuation of applied research, adaptation to changing market needs and creation of variants	• The necessity of persistent research activities along with a successful diffusion process is often underestimated, as producers concentrate on market exploitation and research programmes are regarded as being of secondary importance. The need for ongoing research must be clearly defined	• Continuation of **applied, co-operative research programmes focused on bottlenecks, efficiency gains and variants**
Broad availability of technological knowledge to producers	• Markets are often not developed fast enough since the new knowledge diffuses too slowly or is restricted to a limited number of actors	• **Indirect-specific promotion** of technology diffusion on **demand side** with direct or indirect subsidies (grants, free training and skills, tax advantages etc.)[18] • Support knowledge and technology **transfer** activities, including management of knowledge monitoring and absorption in industry
Regulatory framework conducive to diffusion	• Quick diffusion may be hampered by counterproductive regulations or by competing standards	• **Regulations and standardisation** processes foster variation by securing interoperability • Development and maintenance of an accepted standard (support responsible actors)
Infrastructure and competing technologies	• Depending on the level of novelty of an innovation, there might be a lack of complementary technologies or even of the infrastructure needed for diffusion	• **Infrastructure development**

[18] Grant or tax credit for users until dedicated budget is spent, creating first wave of applications and demand-increasing experiences needed for later dissemination.

Table 3.10 (continued)

System functions and pre-conditions	Potential system failure	Policy instrument to address this failure
Increasing scale needs increasing manufacturing, services, installation and maintenance capabilities for the new technology (also applies to phase 6)	• Producers, service providers, engineers and tradespeople may not be sufficiently capable of producing and applying a new technology, high learning costs have to be borne	• Indirect specific support measures and innovation management programmes in order to **support industry in adjusting to new production needs** • **Education and formation** of engineers, trades etc. needs to be adjusted • Technology **consultancy**
Phase 6: Diffusion		
The system must develop a persistent, self-running market	• There is a tendency to retain demand measures too long, involving the risk of creating dependency on subsidies, cementing early technology paths and preventing adaptation and variation in the diffusion process	• Tailor demand measures and continuous **checks for appropriateness** • Adjustment to the development of niches
The system needs to ensure a constant flow of new knowledge to • close gaps and develop variants • enable the next technological generation (through appropriate financing of (if needed basic) R&D	• Once technologies have diffused broadly in the market, the saturation may be overcome with more variation or with significant improvements of cost-benefit ratios. In many cases this is not achieved through the market reinvestment of producers in R&D, but a real leap forward may require significant efforts in public, more basic oriented R&D	• Keep up **reduced institutional funding** for public research institutes relevant for the improvement of the technology and human capital formation for sustainable development • Activities like the Industrial Research Associations of the German AiF and other industrial working groups meeting and discussing industrial research are needed in order to **enable the feedback loops between diffusion and knowledge creation** (early warning signals, new technological possibilities etc.)

in a naïve or mechanistic fashion. The policy instruments that may fit a certain failure – phase constellation must be applied within a given policy context and political situation. An instrument selected on the basis of this matrix may not fit the existing political culture or the overriding policy trajectories. Still, thinking systematically about appropriate policy instruments based on the matrix should serve to reveal constraining or even dysfunctional trajectories and policy contexts, thereby contributing to a more enlightened policy debate.

3.7 Structure of the Analysis – Overview

This concluding section presents a detailed sketch of the structure of the analysis of the EduaR&D process suggested to the Ministry of Economics and Technology (see Fig. 3.4). It summarises the methodological setting and the policy questions and their interrelations.

The following "recipe" summarises the various stages of the EduaR&D process which is further described in the manual (see Annex 9.1). While steps (i) and (ii) can be regarded as first, successive steps, steps (iii) to (v) contain feedback loops, i.e. the findings here may influence decisions to be taken in previous steps so that the process must be conducted with a very high degree of co-ordination between the various stages.

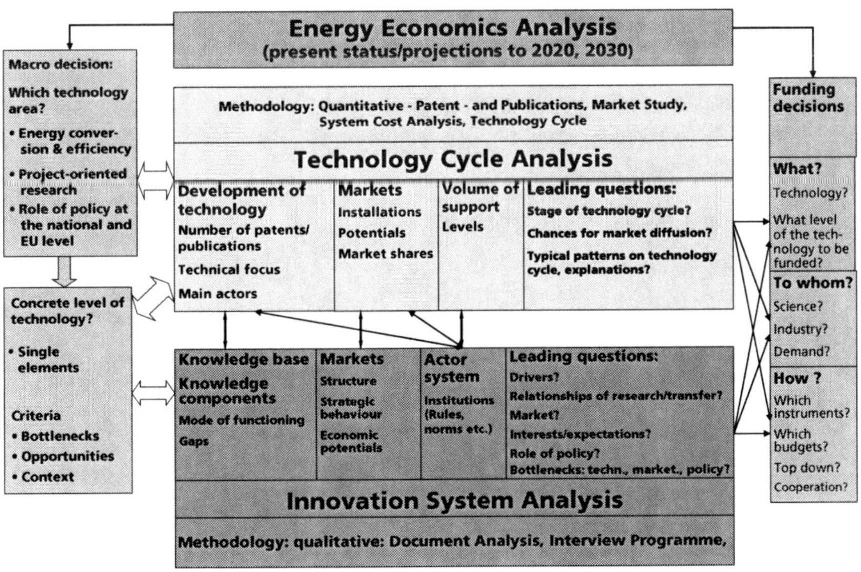

Fig. 3.4 The structure of the suggested analysis for improving the efficiency of energy-related R&D and diffusion policy
Source: Fh-ISI.

i. Preliminary decision with regard to basic energy technology areas and to project-based R&D, politically determined (e.g. by budget portfolio of federal government), supported by brainstorming or interviews with suitable groups.
ii. Energy-economic and climate policy significance (energy-economic benefits, contribution to diversity of energy supply, cost reducing potentials, climate benefits, level of export potential for industry, technological and cost competitiveness of competing energy technologies).
iii. Preliminary decision with regard to technology level of the energy technology considered and – possibly – of its technological competitor:
    a. Technical difficulties, obstacles, market imperfections ("bottlenecks")/ funding leverage
    b. Policy reasons (e.g. priorities, continuation, risks, visibility)
    c. Pragmatic reasons.
iv. Diffusion analysis:
    a. Patent analysis
        - Delineation of the classifications for the individual technologies
        - Identification of the biggest applicants, development over time
        - Patent processes in total (in respective database) and German applicants, if necessary for different variants of the technology
        - Co-patents (co-registrations and/or co-inventors, each differentiated by variants).
    b. Publications analysis
        - Delineation of the classifications for the individual knowledge areas which can be assigned to the technologies
        - Publications in total (in the respective database) and German authors (including variants).
    c. Market analyses (status quo, potentials).
v. Innovation system analysis:
    a. The knowledge dimension: what is the technology and what are its knowledge components?
    b. The economic dimension:
        - Theoretical considerations (type of market etc.?)
        - Existing sector studies (pay attention to definition)
        - Core statements of the energy-economic significance.
    c. Definition of the most important actor groups in the innovation system approach
        - Document analyses
        - Cross-check with the results of the patent analysis.

d. Definition of the most important components (laws etc.) including *funding policy*

- Explorative considerations of the most important relationships
- Interview programme: specific lists of questions for all actor groups

  - Own activities
  - Description of the market and system from the perspective of the interviewee
  - Co-operations and interactions
  - Interests and expectations (technological/economic potentials and expectation of own role etc.)
  - Role of the research sponsorship
  - Bottlenecks

    technological
    market-related
    systemic (cooperation, interaction)

  - Influence on policy/dependence of policy and funding
  - Policy development in the future (desired).

vi. Overall analysis and recommendations for sponsorship policy: horizontal analysis of the interviews, definition of bottlenecks and funding leverage based on diffusion analysis, analysis of potentials and interviews. The intersection of these reveals concrete areas and sponsorship instruments.

References

Arrow, K. J. (1962). The Economic Implications of Learning by Doing, *Review of Economic Studies, 29*(3) June, 155–173.

Baunetz-Online. (2005). *des Zentrums für Umweltbewusstes Bauen e.V. Kassel: Was ist ein Niedrigenergiehaus?* (http://www.baunetz.de)

Bühring, A., Leuchtner, J., Krug, P., & Schüle, R. (2004). Marktpotenzial für Passivhäuser und 3-Liter-Häuser. Fraunhofer Institut Solare Energiesysteme (ISE), Büro für Solarmarketing, Energieagentur Regio Freiburg.

Callon, M. (2001). From science as an economic activity to socioeconomics of scientific research: The dynamics of emergent and consolidated techno-economic networks. In P. Mirowski (Ed.), *Science bought and sold: Essays in the economics of science*. Chicago: The University of Chicago Press.

Carlsson, B., Jacobsson, S., Holmén, M., & Rickne, A. (2002). Innovation systems: analytical and methodological issues; in Research Policy 31 (2002), pp. 233–245.

Chief Scientific Adviser's Energy Research Review Group. (2005). Recommendations to Inform the Performance and Innovation Unit's Energy Policy Review. Report of the Group. London: UK Office of Science and Technology.

Christensen, C. (1997). *The Innovators Dilemma*. Cambridge: Harvard University Press.

Committee on Prospective Benefits of DOE's Energy Efficiency and Fossil Fuel Energy R&D Programs. (2005). *Prospective Evaluation of the Applied Energy Research and the Development at*

DOE (Phase one) – A first look forward. National Research Council. The National Academies Press, Washington D.C. www.nap.edu

Cordis. (2005). Euroabstracts 4/05: Assessing the Potential of CO_2 Capture and Storage (http://aoi.cordis.lu).

Cremer, C. (2005). Integrating Regional Aspects in Modelling of Electricity Generation – The Example of CO_2 Capture and Storage. Zurich: Dissertation-ETH.

Dosi, G. (1988). Sources, procedures and micro-economic effects of innovation. *Journal of Economic Literature, XXVI*, 1120–71.

Dreher, C. (1997). Technologiepolitik und Technikdiffusion. Baden-Baden: Nomos.

Dreher, C., Frietsch, R., Hemer, J., & Schmoch; U. (2006). Die Beschleunigung von Innovationszyklen und die Rolle der Fraunhofer-Gesellschaft. In: Hans-Jörg Bullinger (Hrsg.); *Fokus Innovation - Kräfte bündeln - Prozesse beschleunigen.* 275–306. München/Wien: Carl Hanser Verlag.

Edler, J. (2006). Public action to spur innovation via demand (working title). In P. Smits, S. Kuhlmann, P. Shpaira (Eds.), *The Co-Evolution of Innovation Policy – Innovation Policy Dynamics, Systems, and Governance* (forthcoming).

Edquist, C. (1997). *Systems of Innovation Technologies, Institutions and Organisations.* London/Washington: Printer.

Enquête-Kommission des Deutschen Bundestages "Nachhaltige Energieversorgung unter den Bedingungen der Globalisierung und der Liberalisierung". (2002). Endbericht und Einzelberichte, Drucksache des Deutschen Bundestages 14/9400 vom 07.07.2002, Bundesanzeiger Verlagsgesellschaft, Berlin.

EWI/Prognos. (2005). Energiereport IV. Die Entwicklung der Energiemärkte bis zum Jahre 2030. Energiewirtschaftliche Referenzprognose. Köln/Basel: Prognos AG.

Feist, W. (2005). *Was ist ein Passivhaus?* (Passivhaus Institut http://www.passiv.de) sowie http://home.t-online.de/home/friedl.werner/passiv/passiv1.thtm

Fischedick, M., & Nitsch, J. (2002). Langfristszenarien für eine nachhaltige Energienutzung in Deutschland. Umweltbundesamt Berlin (2002).

Fraunhofer ISI. (1998). *Zwischenbilanz zur rationellen Energienutzung bei Thermoprozessanlagen, insbesondere Industrieöfen.* Karlsruhe: Fraunhofer ISI.

Fraunhofer-ISI, FfE (Forschungsstelle für Energiewirtschaft e.V.) (2003). Möglichkeiten, Potenziale, Hemmnisse und Instrumente zur Senkung des Energieverbrauchs branchenübergreifender Techniken in den Bereichen Industrie und Kleinverbrauch. Karlsruhe: Fraunhofer ISI.

Geiger, B., Nickel, M., & Wittke, F. (2005). Energieverbrauch in Deutschland, BWK Bd. 57 2005 Nr. $\frac{1}{2}$

Gold, B. (1981). Technological diffusion in industry: Research needs and shortcomings. *Journal of Industrial Economics, XXIX*(3), 505–516.

Grupp, H. (1998). *Foundations of the Economics of Innovation – Theory, Measurement and Practice.* Celtenham/Northhampton: Edward Elgar.

Hekkert. M. P., Suurs, R. A. A., Kuhlmann, S., & Smits, R. E. H. M. (2006). Functions of Innovation Systems: A new approach for analysing technological change. In *Technological Forecasting and Social Change* (forthcoming).

Hinze, S., & Schmoch, U. (2004). Opening the black box. Analytical approaches and their impact on the outcome of statistical patent analyses. In H. Moed, W. Glänzel, & U. Schmoch (Eds.), Handbook of Quantative Science and Technology Research. The Use of Publication and Patent Statistics in Studies of the R&D System, 215–235. Dordrecht/Norwell/New York/London: Kluwer.

International Energy Agency IEA. (2004). *Prospects for CO2 Capture and Storage.* Paris: OECD.

International Energy Agency IEA. (2005). *CO_2 Emissions from Fuel Combustion 1971–2003.* Paris: OECD.

IPCC Intergovernmental Panel on Climate Change. (2005). *Special Report Carbon Dioxide Capture and Storage.* Cambridge: Cambridge University Press.

Kleemann, M. et al. (2000). Die Entwicklung des Wärmemarktes für den Gebäudesektor bis 2050. Schriften des Forschungszentrums Jülich, Reihe Umwelt, Band 23

Krewitt, W., Pehnt, M., Fischedick, M., & Temming, H. (2004): Brennstoffzellen in der Kraft-Wärme-Kopplung-Ökobilanzen, Szenarien, Marktpotenziale. Berlin 2004 (in Energy Policy-Article in Press (Science Direct): Krewitt, W., J. Nitsch, M. Fischedick, M. Pehnt, H. Temming: Market Perspectives of stationary fuel cells in a sustainable energy supply system-long-term scenarios for Germany)

Meyer-Krahmer, F., & Dreher, C. (2004). Neuere Betrachtungen zu Technikzyklen und Implikationen für die Fraunhofer Gesellschaft, In: Spath, Dieter (ed.). *Forschungs- und Technologiemanagement: Potentiale nutzen – Zukunft gestalten*, Festschrift für Prof. H-J. Bullinger zum 60.ten Geburtstag. 27–35. München/Wien: Hanser Verlag.

Pehnt, M., & Traube, K. (2004). *Zwischen Euphorie und Ernüchterung. Stand und mittelfristige Perspektiven stationärer Brennstoffzellen*. Berlin: Bundesverband Kraft-Wärme-Kopplung.

Radgen, P. (ed.), (1998). Zwischenbilanz zur rationellen Energienutzung bei Thermoprozeßanlagen, insbesondere Industrieöfen. Karlsruhe: Fraunhofer ISI.

Radgen, P., & Jochem, E. (Hrsg.). (1999). Zwischenbilanz zur rationellen Energienutzung bei Thermoprozeßanlagen, insbesondere Industrieöfen. Abschlußbericht : Zusätzlich mit CD-ROM (Inhalt: Literatur-, Patent-, Förderdatenbank Industrieöfen sowie vollständiger Abschlußbericht als PDF-Datei für adobe Acrobat Reader). Karlsruhe, 1998

Radgen, P., Cremer, C., & Gerling, P., et al. (2005). *Assessment of methods for CO2 capture and storage* (Summary). Karlsruhe, Fraunhofer ISI.

RWE. (2006). http://www.rwe.com.

Schibany, A., Edler, J., Jörg, L., Nones, B, Schmidmayer, J.,Warta, K., & Sheikh, S. (2005). *Evaluierung der Christian Doppler Gesellschaft*. Graz: Johaneum Research.

Schmid, Ch. (2003). *Energiequerschnittstechniken – Einsparpotentiale, Hemmnisse und Maßnahmen zu ihrer Erschließung. In: Rationelle Energienutzung in der Industrie – ein Beitrag zum Klimaschutz*. Düsseldorf: VDI-Verlag.

Sundermann, C. (1999). Constructive Technology Assessment, in: Stephan Bröchler, Georg Simonis, Karsten Sundermann (Hg.): Handbuch Technikfolgenabschätzung, Berlin 1999, edition sigma, S. 119–128

VDEW. (2006). *Stromwirtschaft nutzt vielfältigen Energiemix* (www.strom.de).

Chapter 4
Fuel Cell Technologies

Substantial attention has been paid to fuel cell technology both in R&D and niche markets for almost two decades now. For certain mass applications of fuel cells, very fast market penetration has been projected several times during the last 10 years (Frost & Sullivan, 2001; Wurster, 1999; AMCG, 2002; Wengel & Schirrmeister, 2000); however, these projections did not materialise due to unresolved technical issues and high risks for the investments in mass produced elements of the fuel cell.

4.1 Technology Description and Energy-Economic Significance

Fuel cell technology is often considered an essential element of a sustainable energy system because of its high electricity to heat ratio and high overall conversion efficiency. In addition, the fuel cell's exhaust produces zero emissions when using hydrogen which could be generated by electricity from renewables in the long term or from bio-based methane or liquid bio-fuels. As the demand for heat and fuels is expected to decrease much more than the demand for electricity in the final energy sectors, fuel cell technology is considered a better match for the increasing share of electricity in total energy demand. The criticism made of its technological rivals, i.e. the engine driven cogeneration plant, the micro-turbine or the Stirling engine, is the low electricity share in their total energy output (Krewitt, Pehnt, Fischedick, & Temming, 2004; Krewitt, Nitsch, Fischedick, Pehnt, & Temming, 2006).

The estimated figures on the market potentials of fuel cells until the middle of this century vary substantially: some authors expect a high utilisation of stationary and mobile fuel cell technology (e.g. stationary applications in Germany by 2050: 6,000 MW_{el} installed capacity for district heating, 2,000 MW_{el} in residential buildings and again 6,000 MW_{el} for industrial process heat, Krewitt et al., 2006) and the majority of cars to be powered by PEM fuel cells at an installed capacity of more than 100 GW_{el}. Other authors argue that the stationary use of fuel cells powered by hydrogen based on renewables would not make much sense since the fuel or the electricity from renewables could be used directly, avoiding the reforming or electrolysis process necessary for generating hydrogen. Only the use of mobile

E. Jochem (ed.), *Improving the Efficiency of R&D and the Market Diffusion of Energy Technologies*, DOI 10.1007/978-3-7908-2154-3_4, © Physica-Verlag Heidelberg 2009

fuel cells would make sense in the long term, i.e. in road and ship transportation (Jochem, Landwehr, Tönsing, Nöltscher, & Vollmar, 1997). This conclusion is particularly relevant in the case of high CO_2 reduction targets because hydrogen from fossil sources is either too carbon-intensive or too costly due to carbon capture and storage.

In Germany, R&D is presently being carried out on three different fuel cell systems: the PEM (Proton Exchange Membrane) fuel cell with hydrogen, methanol or reformed natural gas as fuels, the MCFC (Molten Carbonate Fuel Cell) and the SOFC (Solid Oxide Fuel Cell) using gaseous or liquid hydrocarbons. Alongside the actual fuel cell stack, other components are required for the operation of the fuel cell system such as e.g. reformers for hydrogen production, as well as measurement and control systems to control the total operation and storage for the energy source.

In principle, all the systems are currently only available as prototypes; the energy-economic significance is based on expectations of future markets. The scenarios developed by the Brennstoffzellen-Bündnis Deutschland (2004, Fuel Cell Association of Germany) for individual applications are:

- Industry: 2006 approx. 15 MW; 2010 approx. 300 MW; 2015 approx. 1.3 GW p. a. (MCFC, SOFC).
- Residential: in the period 2008–2010 several thousand systems, in the period 2011 up to 2015 several ten thousands for households (PEM, SOFC).
- Mobile: extension of fleet vehicles within the scope of the EU Lighthouse projects to commercial fleets and a broad market introduction from 2015 (PEM).
- Portable: commercial application of some products in niche and luxury markets to start with, but then in a broad mass market by 2010 (DMFC, PEM).

The key question for further market potential mobilization is a steady decrease in fuel cell costs and increase in lifetime (see Table 4.1). Moreover, future energy relevance of CHP systems and fuel cell systems is heavily dependent on political incentives and the creation of a more decentralized and efficiency-oriented electricity

Table 4.1 Cost degression assumptions for selected fuel cell systems (€/kW), projected for 2010–2030

	kW (net)	Life time (a)	2010	2020[19]	2030
PEMFC	5	15	1500	1000	750
PAFC	200	15	1625	1355	1080
MCFC	500	15	1250	1000	1000
SOFC	2000	15	1050	875	750

Specific investment costs, not incl. interest payments for construction (Bauzinsen)
Source: Enquête-Commission, 2002

[19]The National Development Plan H2 and FC Technologies Version 1.1 of the Strategy Council 25 July 2006 gives higher figures for 2020: 1,700 Euro/kW for energy supply in households and 1,000 to 1,500 Euro/kW for stationary industrial applications.

Table 4.2 Systems and key components of fuel cell technology

Systems	Key components
PEM	Membrane, bipolar plates, gas diffusion layers, design
Reformer	Steam reforming, partial oxidation and autothermal reforming
MCFC	Electrodes, electrolyte, design
SOFC	Electrodes, electrolyte, planar and tubular design
Measurement and control systems	Inverted rectifiers, sensors, pumps
Storage of hydrogen	Cryogenic storage, pressured storage, metal hydride, chemical hydride

supply structure. The substitution of up to nearly half of total power plant capacity in the next five decades is the maximum potential.

This analysis aimed to cover all three major fuel cell systems in Germany. However, the early steps of this analysis made it clear that the PEM variant was the most promising, so there was a focus on this variant. Furthermore, rather than analysing the fuel cells as such, key components were identified that appear to be crucial for the further development of fuel cell markets in Germany (see Table 4.2, see also Sect. 4.2.1 for a description of the individual components).

4.2 The Analysis on the Micro-Level

4.2.1 Technological and Economic Challenges

This section describes the major technological – and related economic – challenges that fuel cells still face today. Since fuel cells represent a cross-cutting energy conversion technology, it was not clear from the outset at which level the analysis would be most effective. If the analysis covers a large technological domain, there is the danger of only being able to make very general conclusions without much value for concrete policy decision making. On the other hand, if the analysis is reduced to a very narrow technology level, there is the danger of no longer being able to refer to a statistical basis for many indicators within the technology cycle analysis, which then gives rise to non-representative individual results.

The statements here are based on literature and interviews. As a rule, the PEM was analysed separately in the analyses because, as mentioned above, this fuel cell variant seems to be the most promising. Where this made sense, analyses were also conducted for other individual components or a search was made for "fuel cell" cell (e.g. publications in popular scientific literature).

Overall, it appears that research and development across all the fuel cell variants must concentrate on the operating life and costs. This is not only valid for stacks but also for the periphery and the integration of all systems.

4.2.1.1 PEM – Fuel Cells

The electrolyte in the polymer electrolyte membrane fuel cells (also called PEMFC or SPFC: solid polymer fuel cell) consists of a thin, alkaline ion exchange membrane which acts as a proton conductor. The working temperature of the PEM is approx. 50–60°C, but also as high as 190° if polybenzimidazole is used as the membrane. Other cost-intensive parts of the PEM include the bipolar plates (approx. 30%) and the gas diffusion layer (15%). Besides cost reduction, the objective here is to improve the operating life.

4.2.1.2 Reformer

Basically the three methods suitable for reforming hydrocarbons for FC systems are also those which are important in big chemical plants. Steam reforming, in which hydrogen is produced in an endothermal reaction by the catalytic conversion of hydrocarbons or alcohols with steam; partial oxidation in which the fuel is reacted exothermally with a sub-stoichiometric amount of oxygen; and autothermal reforming, which combines catalytic partial oxidation and reforming in one reactor. After reforming, there is a further stage of CO conversion. These systems need to be more compact and efficient.

4.2.1.3 MCFC

The electrolyte is usually a combination of alkali carbonates which are retained in a ceramic matrix. The operating temperature is approx. 650 C. The electrodes are usually made of nickel or nickel oxide. In order to obtain the desired performance at lower temperatures, on the one hand, the catalyst's activity has to be improved and, on the other, the corrosion of the bipolar plates and the carbonate degradation reduced.

4.2.1.4 SOFC

The oxide ceramic fuel cell (solid oxide fuel cell, SOFC) works between 800 and 1000°C. The electrolyte of this type of fuel cell consists of a ceramic able to conduct oxygen ions (yttrium-stabilised zirconium oxide, YSZ). The electrodes are usually made of Sr-doped $LaMnO_3$ as the cathode and a nickel/YSZ metal ceramic. The main field of application is combined heat and power in larger stationary installations. Alongside this, there are applications in smaller block heating stations (Sulzer-Hexis) and as battery substitutes (Auxiliary Power Units) in cars (BMW). Two SOFC designs are currently being researched which differ in their structure (planar, tubular). In the planar cells, the main aim is to reduce the temperature so that cheaper materials can be used and the lifetime extended. In tubular systems, the power density is to be improved by better cell geometry (flat, compact tubes).

4.2.1.5 Measurement and Control Technology for Fuel Cells

The further the core components of the fuel cell system are developed, the clearer the significance of the periphery for the supply with process gases and fluids, heat conduction and the conversion of electricity and heat to the desired form. Development goals here include the electrical and thermal integration of the different units and the reduction of parasitic energy consumption.

4.2.1.6 Hydrogen Storage

Different storage technologies can be used depending on the power of the FC system and the amount of the secondary energy source to be stored. For large amounts, these are the cryogenic storage of hydrogen or pressured storage. For smaller amounts, standard gas cylinders can be used. For storing very small amounts of hydrogen, hybrid storage using reversible metal hydrides such as $Ti(Cr1\text{-}xMnx)_2$ can be used. The storage should have the highest possible energy density. In stationary systems, as a rule natural gas is used.

4.2.2 Position in the Technology Cycle

This section analyses the position of the various technologies in the technology innovation cycle.[20] By applying various sorts of indicator-based analyses, it tries to show how mature the technologies are, how much and what kind of R&D is still being done and how near to market introduction the technologies are.

4.2.2.1 Research Activities – Patent and Publication Analysis

We start with a patent and publication analysis for the technology field "fuel cells", whereby patents indicate market-oriented research and publications indicate more basic-oriented research activities (see Fig. 4.1). The Science Citation Index and the database of the European Patent Office were used as sources.[21] Overall, there is marked growth, albeit with varying intensity. Compared to the overall trend for all patents at the European Patent Office, the increase during the last 15 years is much higher for fuel cells (see Fig. 4.2, "Trend"), with the exception of the MCFC patents.

The significance of the PEM is reflected in the number of patents for this variant which account for approx. 80% of the patents and approx. 45% of the publications on fuel cells in 2003 (see Fig. 4.2). It is thus obvious that the broadest research is

---

[20] An integrated dynamic assessment of CHP with fuel cell technologies is being made in ongoing EduaR&D projects. The projects are carried out by the Research Center Juelich and the Research Institute for Energy Economics (FFE). The results can be used as a further indicator for the position in the technology cycle (Riensche, 2006).

[21] Database provider Questel/Orbit – Intellectual Property Group, database EPPATENT or database provider STN for the publication analysis.

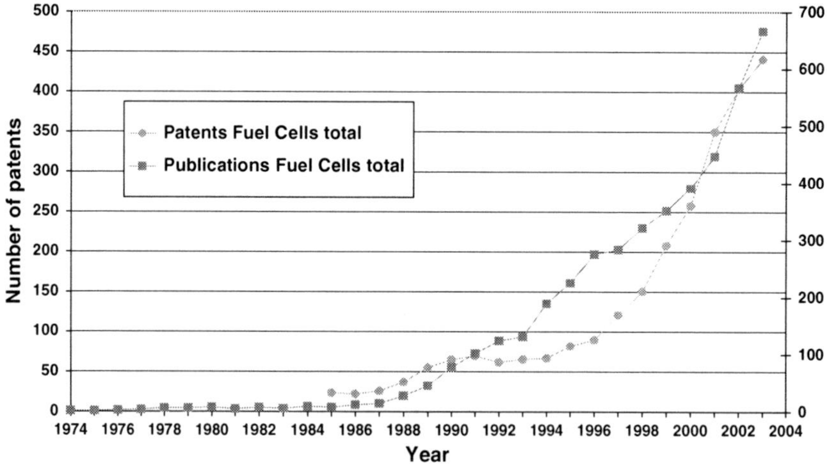

Fig. 4.1 Patents (EPO) and publications (SCI) on fuel cells, 1974–2003

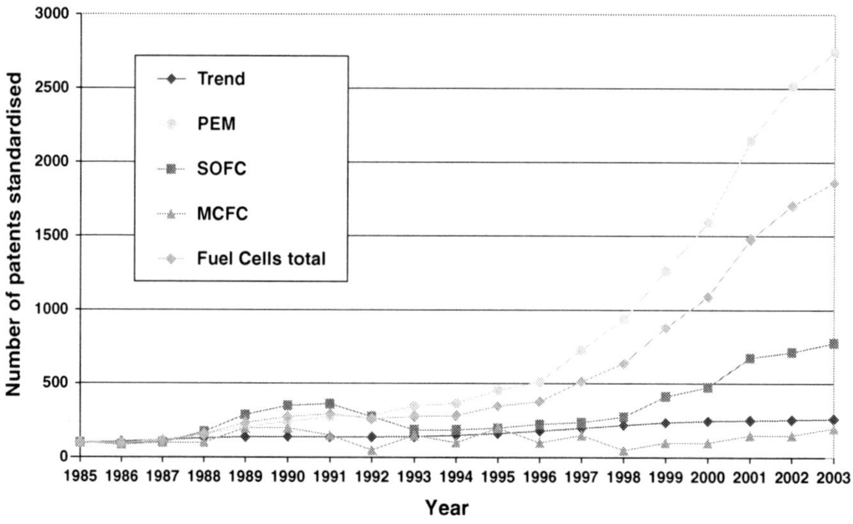

Fig. 4.2 Comparison of the fuel cell patent development with general patent development (1985=100) at the European Patent Office, 1985–2003

taking place in the field of PEM. It is also of interest, in turn, that the publications continue to increase substantially accompanied by more intensive patent activities (see Fig. 4.3). This shows that (Fig. 4.4) many scientific questions are still unresolved in the field of PEM and SOFC or that technical improvements based on new scientific research results can be expected in the future. The knowledge base for the technology is obviously not yet mature. More accurate analyses of this phenomenon and thus a delineation of the actual technological bottlenecks or opportunities would

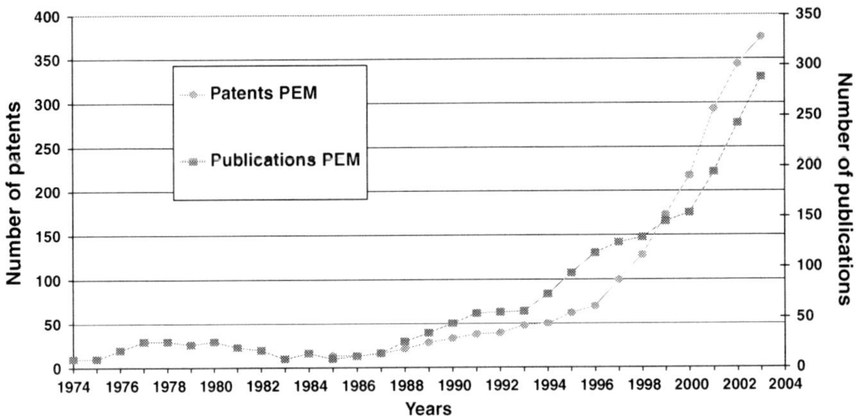

Fig. 4.3 Number of PEM patent (EPO) and publications (SCI), 1974–2003

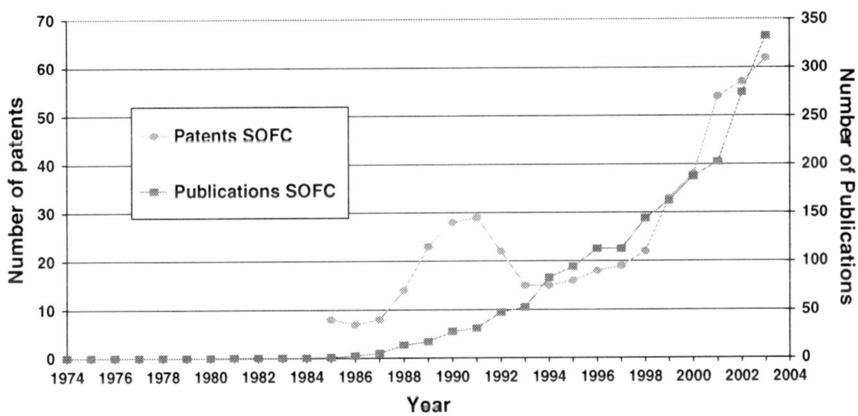

Fig. 4.4 Number of SOFC patents (EPO) and publications (SCI), 1974–2003

require more detailed analyses on the level of individual components. This should be tackled in future in a further development of this methodological approach.

The patterns for SOFC are very similar, with one exception: while the number of SOFC publications increased each year, the number of patents clearly peaked in the year 1990/1991. This level was only reached again in 1998 (see Fig. 4.4). In this field, therefore, there is a disenchantment phase in the mid-nineties and then new dynamics appearing from the second half of the nineties. However, as for PEM, it is also valid here that there is an obvious unbroken demand for basic scientific clarification of the facts.

No statement about a trend is possible for MCFC because of the low number of annual patents (three on average). The number of publications (about 50 per year in

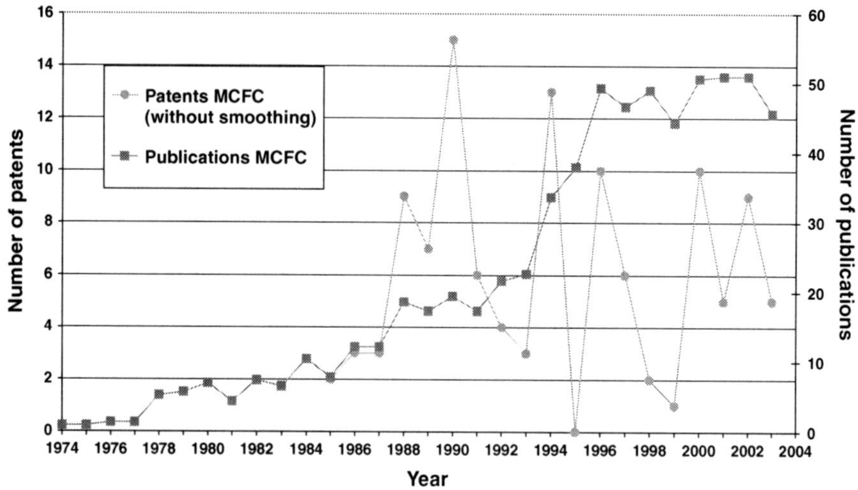

Fig. 4.5 Number of MCFC patents and publications, 1974–2003

the last 10 years) is also much lower compared with PEM and SOFC and has been stagnating since 1996 (see Fig. 4.5).[22]

Looking at German applications at the EPO, we observe an even higher predominance of PEM patents in Germany, albeit with a slight decrease since 2001, in marked contrast to the international trend. The number of patents for SOFC and MCFC has remained at the same level for several years indicating a lower activity level for SOFC in Germany compared to PEM (see Fig. 4.6).

The share of German publications in the field of fuel cells appears to be a bit lower than the share of patents[23]. (see Fig. 4.7). Whereas there has been remarkable growth in the numbers of PEM and SOFC publications since 1993, the number of MCFC publications increased only slightly, remaining at a much lower level of some 4 publications per year compared to 25–30 publications for the two other types during the last few years. The peak, already shown in Fig. 4.4 for SOFC patents, can be seen again for publications, although not as marked.

In order to better understand the major drivers and bottlenecks, more detailed patent analyses were carried out for the key components of PEM described in Sect. 4.1. For these components, a search was made using key words and various combinations of classifications alongside patent classifications, and the results

[22]The search strategy for patents was modified to address a comment of MTU that the number of patents in this figure is too low. On one hand, the patent classes IPC H01M-008 and H01M-004/88 were used with the keywords "Molten Carbonate or Liquid Electrolyte" in the abstracts, on the other hand all EP patents of MTU and IHI with the keyword "fuel cell" were taken into account.

[23]A lower representation of German publications in the American SCI database can be assumed; however, analysis made in alternative databases (Compendex, Pascal) did not yield a different result.

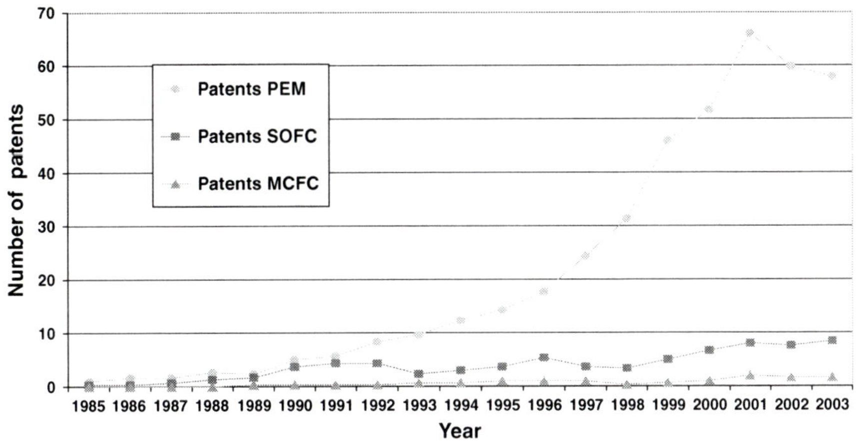

Fig. 4.6 Number of patent applications from Germany at the EPO, 1985–2003

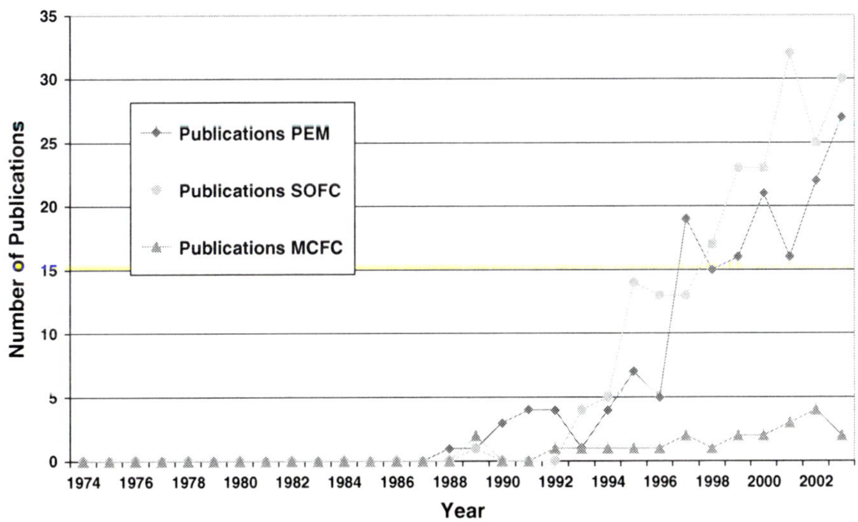

Fig. 4.7 Number of German publications on the three types fuel cells in the Science Citation Index SCI, 1988–2003

obtained were sorted by applicant. The resulting graphs showed a steady increase until 2002 and offer evidence of unbroken R&D on the individual components (see Fig. 4.8). The overall importance of the membrane is striking. Furthermore, the numbers of annual applications for bipolar plates (BPP) and the gas diffusion layer (GDL) decrease after 2002, while the increase in membrane applications flattens out. This correlates with the equally lower number of fuel cell patents in the field of mobile applications.

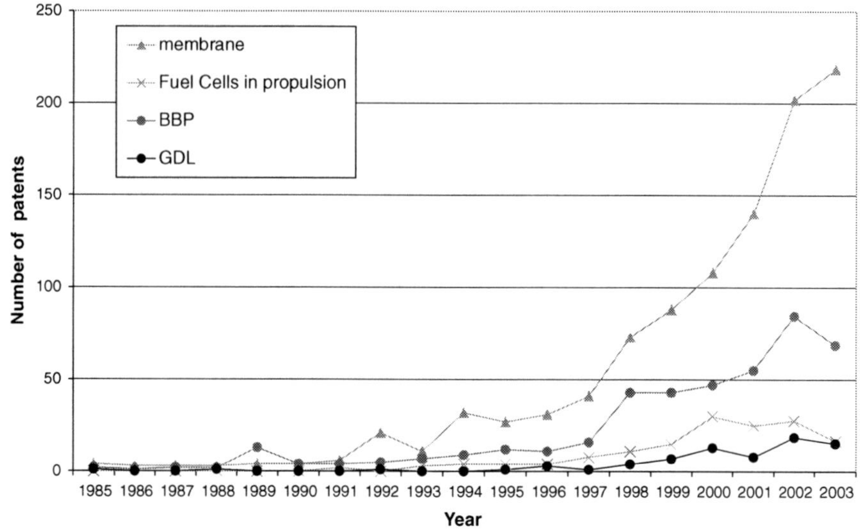

Fig. 4.8 Patent applications for components of the PEM fuel cell, 1985–2003

Furthermore, for the PEM technology, the origin and type of applicants were analysed more closely (see Figs. 4.9 and 4.10). According to this, Germany, together with the USA, is among the most important applicants at the European Patent Office after Japan. The German share was actually the highest in the year 2000 with approx. 200 patents and then falls to approx. 150 in 2001.[24] Interestingly, until 1990 almost all patents were produced by companies, while today the applicants also include research institutes and, since the end of the nineties, an increasing number of (see Fig. 4.10).

By analysing citations of previous patents in patent applications one can obtain an indication of how important certain application years were for subsequent periods. As a rule of thumb, the more subsequent patents cite previous ones, the more they build on previous knowledge. The less they cite previous patents, the more diverse or novel the research carried out. Within the period under observation, the number of cited patents in the patents of that year and the subsequent four years fell (see Fig. 4.11). Whereas the patents from the year 1988 were each still cited twice in the subsequent period of four years, less than half the patents from the years 1998 to 2002 were cited at all. This result indicates, on the one hand, the importance of research carried out in the mid and late 1980s for subsequent periods. On the other hand, it shows that research developed dynamically and took new directions during the 1990s.

[24]This trend can also be observed in the SOFC applications. The respective number of German patents differs in the pure frequency count (Fig. 4.7) and in the analysis of the country of origin. This is due to the fact that, with the same underlying number of patents in both evaluations, the analysis of country of origin counts each inventor in a patent.

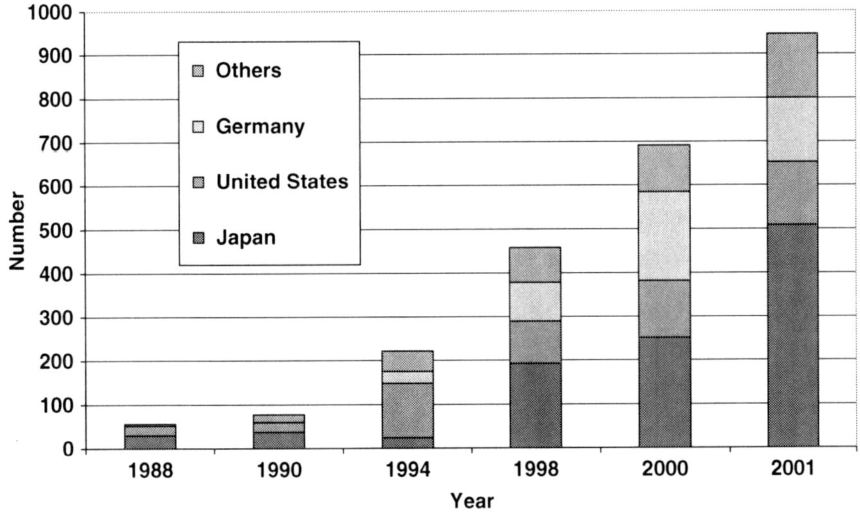

Fig. 4.9 Number of PEM patents by country of origin (EPO), 1988–2001

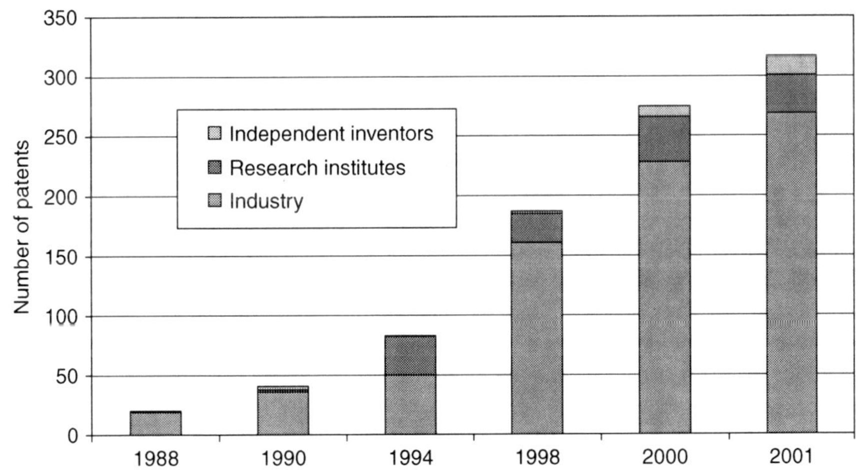

Fig. 4.10 Number and type of applicants for PEM patents, (EPO), 1988–2001

4.2.2.2 Markets

It is clear that a broad mass market is not expected until after 2010 (Gummert & Suttor, 2006). But worldwide, there are a number of pilot and demonstration systems already installed in cars, portable applications and stationary plants. It is very difficult to get a complete overview of installations because, on the one hand, different players such as utilities, manufacturers or users publish data for "their" fuel cell, and, on the other hand, if an installation does not work well, no data are published

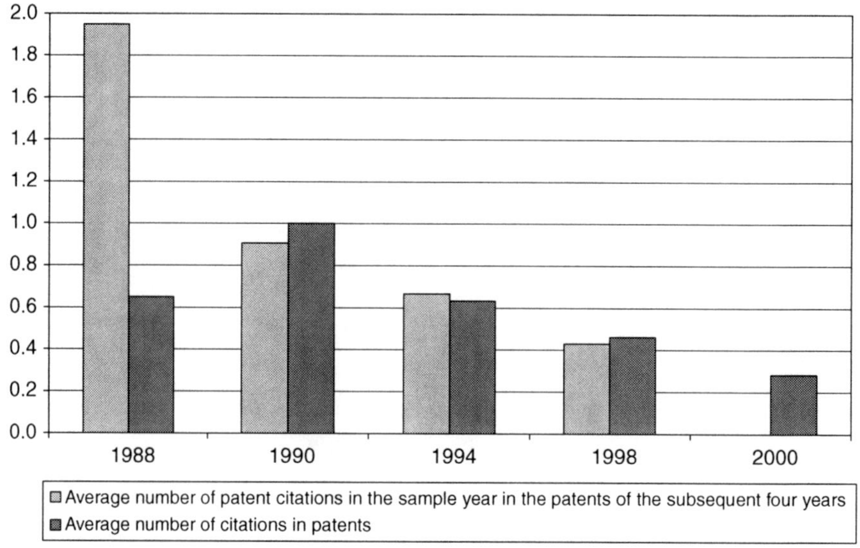

Fig. 4.11 Number of patent citations in fuel cell patents, 1988–2000

or sometimes the system is shut down. Nevertheless, Fuel Cells 2000 have set up a fuel cell database for stationary applications[25] which has approx. 900 entries.

Obviously one can distinguish between different time phases: for example, in the 1990s, mainly PAFC systems were built. The AFC had its field day during the nineties,[26] only a dozen of them were installed in 2001, but since then they have appeared more often again up to the present. UTC is the largest manufacturer of AFC (with 183 of the 230 global AFC installations).[27] The PEM entered the market in 1988 as the most promising technology and the number of installations continued to grow up to 2004 (see Fig. 4.12); the numbers for 2005 may not be complete).

Comparing the strengths and weaknesses of the different variants, it appears that German industry has abandoned the race for micro-cells for consumer electronics (4C application) based on PEM technology. Nor does it seem likely that it will be able to catch up again, given the lack of companies investing in this area and the lack of active public research institutes. There are German actors driving other PEM developments such as vehicles (most importantly Daimler Chrysler and Opel/Vauxhall) and CHP, although Japanese and American producers are still in the lead.

As for system providers, there are a few promising examples of system developers who have already created a market for themselves, such as low temperature

[25] http://www.fuelcells.org/db/

[26] 4 installations in the 1980s and 150 in 1990s.

[27] UTC originally wanted to make a complete switch to the PEM technology, but is now once again backing the proven AFC technology since the targets for PEM have not been achieved.

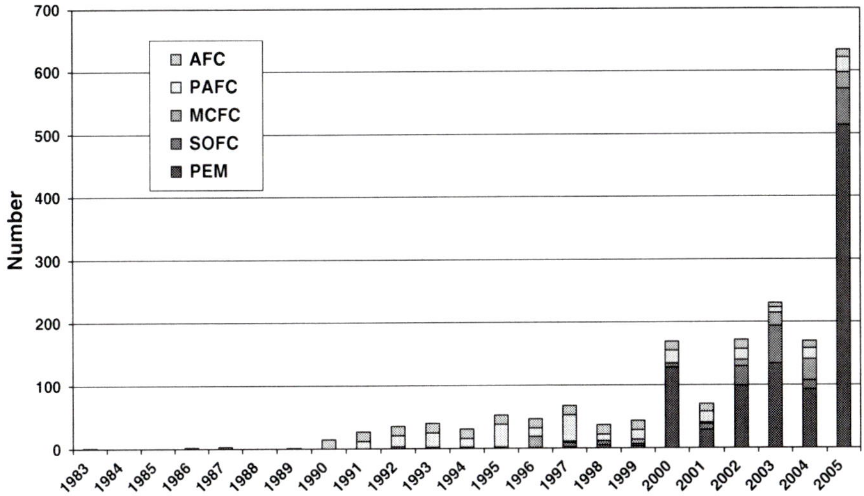

Fig. 4.12 World-wide installations of stationary fuelcells 1990–2005
Source: http://www.fuelcells.org/db/ as of January 2005; FDIC; Fraunhofer ISI

DMFC applications, e.g. in caravans, boats, portable electronic devices or a few industrial applications in the area of traffic and environment measurement technologies or Siemens/HDW with submarines. Other system providers which are still at the level of demonstration projects or small niche markets in Germany include RWE Fuel Cells (stationary fuel cell for decentralised energy production), Siemens (small SOFC power plants), Vaillant, efc and Viessmann (heating).

While publications and patents indicate technological improvements, one indicator for the potential *market development* is the number of company start-ups.[28] According to commercial registrations (see Fig. 4.13), there was a fivefold increase in the numbers of companies in the field of fuel cells in the period 2000/2004 compared with 1995/1999. This figure only hints at the company development: The method is deficient in that fuel cell companies' insolvencies are not registered.

The market for fuel cells in Germany, especially regarding the PEM-driven vehicle application, has been characterised by several waves of pre-mature euphoria, illustrated by very ambitious predictions of car fleets running on fuel cells. Most of these predictions have not been realised.

An indicator for the importance of fuel cells in the general discussion is the citation of fuel cells in the scientific literature shown in Fig. 4.14 as the sum of seven newspapers/magazines. Here we see stable growth up to 2001, and it is evident that fuel cells underwent a euphoric phase whose expectations were not met and which has resulted in risk aversion among many actors. If this curve is compared

[28] Within the scope of the research project, the attempt was also made to analyse the development of registering trade names for fuel cell products over time. A lack of classification prevents a more precise study – since 1997, however, there has been a clear increase in the registration figures.

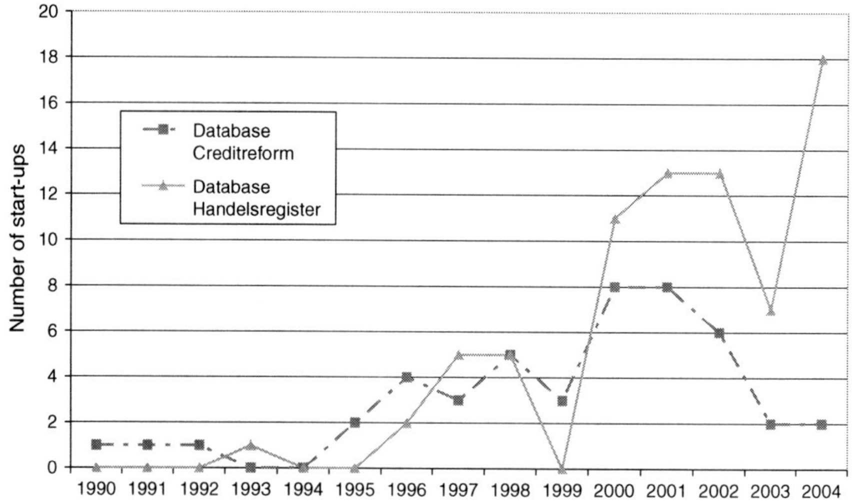

Fig. 4.13 Company start-ups concerned with fuel cell technology in Germany, 1990–2004

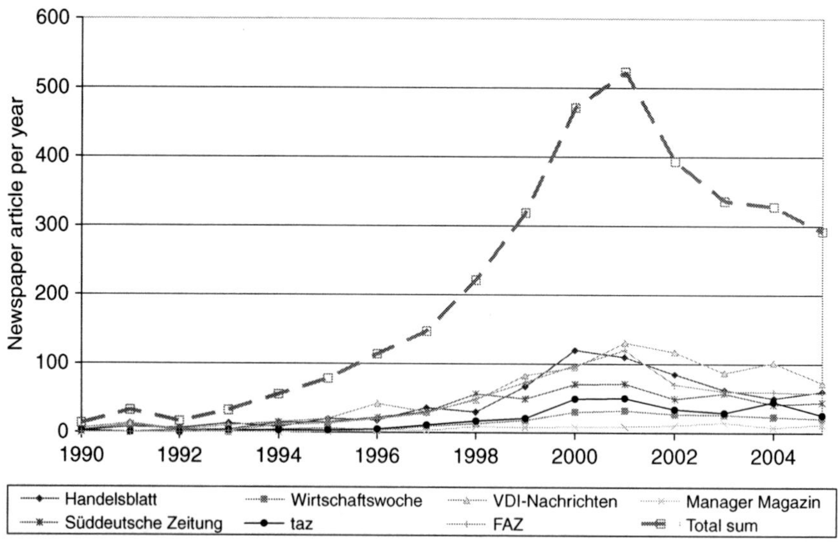

Fig. 4.14 Fuel cell citations in popular literature in Germany, 1990–2005
Source: own analysis

with the public R&D expenditures, then this euphoria slightly precedes the intensive financial incentive programme ZIP[29] (see Sect. 4.3.3.2).

Based on the interviews conducted, a more realistic picture regarding future developments now sees growing niche markets for several applications such as portable PEM applications, and for fuel cell components (e.g. SGL Carbon), but there is still no mass anonymous market for systems. Both R&D services and the producers of key components are still working on a face-to-face basis: suppliers of components know their clients, the number of clients is still modest, fully fledged markets and distribution structures have not yet developed.[30] In many areas, the development is characterised by a high number of demonstration projects rather than self-sufficient markets. This is also indicated by the very low number of brand names used for fuel cell applications and components in Germany, although in some areas, such as mobile applications based on DMFC, there are indications that brand names will become important in the near future.

For example, for PEM application, Siemens has a small, stable and exclusive market for a submarine power system. For PEM applications in vehicles and in heating and power systems (co generation), two further generations will be needed before the technology is efficient enough to support a market in Germany. The number of demonstration projects running in Germany is regarded as very high compared with other European countries.

Only one of the companies interviewed indicated a large number of products sold on an anonymous market and with a wide variety of applications, mainly in the area of on-board electricity, for example in boats and camping vehicles. This niche market appears to be stable already and will most likely continue to grow in the future.

With regard to industry's expectations of future market developments, the scenarios are less ambitious than they were in the past – and none of the interviewed actors rely on commercial market studies. The time horizon for a moderate market take-off is now given as between five years (heating in private households) and 15 years (vehicles). For residential energy, between 5 and 20% are expected in the long term in Germany (while in Japan, for example, the market penetration is predicted to be much higher due to a policy of decentralisation). Expectations are much more modest for mobile applications in vehicles, ranging from 1 to 10% within the next 15 years. For portable systems, the market seems especially promising for computers. Via expensive niche applications at the beginning, a mass market with 25% to 50% penetration is conceivable because computer producers expect their computers to have a much higher power demand which will not be able to be met by traditional accumulators. These data seem to be quite similar to the figures specified by

---

[29] ZIP (Zukunfts-Investitions-Programm): "Programme on investment in the future" – this programme covers R&D in the area of non-nuclear, environmentally friendly energy research, e.g. fuel cells.

[30] As a reminder: the more important figure for the producers of components is not the number of applications but the installed electric power.

Table 4.3 Set of indicators for PEM fuel cells along the technology cycle, status 2006

			1	2	3	4	5	6
Phase			Discovery	Euphoria	Disillusionment	Reorientation	Rise	Diffusion
Patents and publications								
A.1	Number of publications	n	low	***medium***	***medium***	***medium***	***high***	***medium***
A.2	Publication trend	n/a	***increasing***	***increasing***	stagnating	stagnating	***increasing***	stagnating on high level
A.3	Average number of citations from publications	n	*very high*	*high*	*medium*	*medium*	*medium*	*medium*
A.4	Publication trend in popular literature*	n	low	increasing	***stagnating***	***decreasing***	increasing	***stagnating***
A.5	Number of patents	n	low	high	low	medium	high	very high
A.6	Applicants: Enterprise size		***major enterprises***	SME	***major enterprises***	***major enterprises***	***major enterprises n***	SME
A.7	Applicants:		research institute	***industry***	***industry***	***research inst./ind.***	***industry***	***research inst./ind***
A.8	Trend of patents	n/a	***increasing***	***increasing***	decreasing	stagnating	***increasing***	stagnating on high level
A.9	Average number of citations from patents	n	very high	high	***medium***	high	***low***	***low***
A.10	Age of citations in patents	a	***low***	***low***	medium	medium	high	very high
Markets								
B.1	Market penetration	%	***low***	***low***	medium	medium	high	very high
B.2	Cumulated number of users	n	0	***low***	***low***	***low***	medium	very high
B.3	Cumulated number of sold units	n	0	***low***	***low***	***low***	medium	very high
B.4	Number of competitive techniques*		low	***medium/high***	low	1	low	high
B.5	Use of brand names		*no*	*yes*	*no*	*yes*	*yes*	*yes*
B.6	Share of value added*	%	*low*	*low*	*medium*	*medium*	*high*	*very high*
B.7	Number of suppliers	n	0	***medium***	low	very low	***medium***	high
B.8	Number of start-ups*	n	0	***medium***	low	very low	***medium***	high
B.9	Insolvencies	n	*0*	*0*	*many*	*many*	*a few*	*medium*
B.10	Number of outfitters	n	*0*	*medium*	*low*	*very low*	*medium*	*high*

Table 4.3 (continued)

			1	2	3	4	5	6
Phase			Discovery	Euphoria	Disillusionment	Reorientation	Rise	Diffusion
B.11	Number of component suppliers	n	0	***medium***	low	very low	***medium***	high
B.12	Number of specialised suppliers	%	0	***low***	0	very low	***low***	medium
B.13	Type of co-operations		*development partnership*	*development partnership*	*horizontal co-operation*	*vertical co-operation*	*product partnership*	*service/processes*
Frame conditions								
C.1	R+D expenditure	Euro	*low*	*high*	*medium*	*medium*	*high*	*medium*
C.2	Trend of R+D expenditure	Euro/a	*increasing*	*increasing*	*decreasing*	*stagnating*	*increasing*	*stagnating*
C.3	Research promotion	Euro	***medium***	***very high***	***medium***	very low	low	low
C.4	Number of proposals for R+D	n	*low*	*high*	*very high*	*low*	*low*	*medium*
C.5	Classification of agents		*research institute*	*industry*	*industry*	*research inst./ind.*	*industry*	*research inst./ind*
C.6	Investment in technology development	Euro	*low*	*high*	*medium*	*medium*	*high*	*medium*
C.7	Consideration in technological impact assessment	n	*low*	*medium*	*medium*	*low*	*high*	*low*
C.8	Market studies	n	*0*	*many, broad*	*medium*	*medium*	*high*	*medium*

* This indicator was not just applied to PEM but to fuel cells in general.

the Fuel Cell Association of Germany (BZB) given above which is not surprising considering the close contacts between the interviewees and the BZB.

For PEM fuel cell technology each of the six stages within the technology cycle is characterised by a certain set of values for the indicators (see Table 4.3). This is a simplification since the table does not show past developments and the visual representation suggests that each indicator has the same relevance. Even though it should not be over-interpreted, it does give a comprehensive overview of the state of the art of this technology, but could be certainly elaborated further. However, this report focused on the development of a methodology and less on the completeness of data that may be available.

The current state of PEM fuel cell technology is shown by the present levels of indicators marked in bold italics text, e.g. the indicator "patent trend" (A 8) is increasing. Indicators for which data are not available or not significant are highlighted in italics. Thus, the majority of indicators suggest that the PEM fuel cell is still in the (late) euphoric phase. Of course, the PEM researchers and developers would argue that its status is more advanced and shortly before take-off in several niche markets in phase 5.

To conclude, the table demonstrates the difficulties of assigning a given technology to a clear position in the technology cycle. However, it shows the advantage of improving the transparency of the discussion, it suggests more differentiation into the different applications of a technology (here of the PEM fuel cell), and clarifies R&D objectives and policies.

4.3 The Technological Innovation System

4.3.1 The Overall System

The innovation system for fuel cells in Germany consists of a *set of innovation systems* for the different fuel cell variants. These innovation systems overlap and interact, of course. Although the major technological principles are similar, there is a high degree of *specialisation* within the overall fuel cell innovation system. For example, only very few companies produce and sell more than one of the key components, and the producers of fuel cells have tended to specialise in one major application area relying on one major technological variant. Still, the market players and research institutes are well informed about the overall development of fuel cells in Germany, and seem able to assess the interdependencies of the variants' developments.

Even restricted to one technology, it is not easy to capture the major actors of the PEM fuel cell innovation system in Germany (see Fig. 4.15). The PEM innovation system is addressed in line with the overall focus of this analysis on PEM fuel cells and given the complexity of all fuel cell variants. The distribution of actors reveals a rather low number of basic research institutes, an absence of any meaningful technology transfer institute, a fairly large number of component producers and a small number of system producers. The industrial companies involved are all large

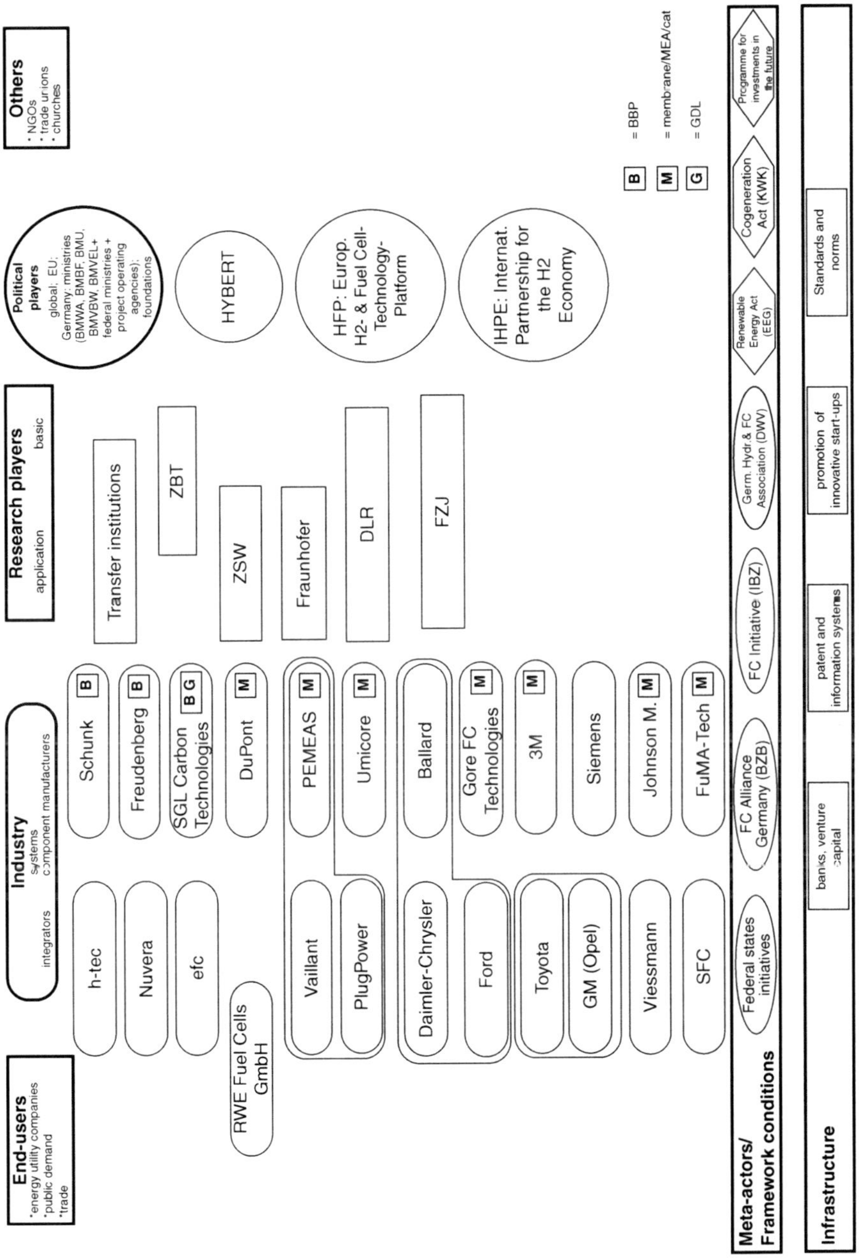

Fig. 4.15 The German innovation system for PEM

enterprises who usually produce fuel cell components as only one part of their portfolio. All the actors shown in the diagram are known to each other and one interviewee talked about “the usual suspects” in the German fuel cell community. These “usual suspects” are organized in numerous alliances, initiatives and competence centres. Basically, each Bundesland has its own organization to foster fuel cell activities in its specific region.

4.3.2 The Various Elements of the Innovation System and Their Roles

4.3.2.1 Industry

Overall, interviews confirm the pre dominance of an oligopolistic market structure for the various technological variants and applications which corresponds to the niche character of the related markets. There are no indications that the number of market players is growing; on the contrary, in some areas – such as heating systems suppliers– the market has consolidated in the last few years.

The production of *components* such as *BPP, GDL* and *MEA* (membrane electrode assemblies) is one strength of the German system; some of the component producers cover the whole value chain and are rather autonomous in the development of their specific component. Only in one area is there a shortage of component providers, i.e. for DMFC in mobile applications, where a market is already established and greater quantities are needed.

As for system providers, a few promising examples have developed a market for themselves already, such as SFC with a considerable market in mobile, low temperature DMFC applications, e.g. in caravans, boats, portable electronic devices, a few industrial applications in the area of traffic and environment measurement technologies or Siemens/HDW with submarines. Other system providers who are still at the level of demonstration projects or small niche markets in Germany include RWE Fuel Cells (stationary fuel cell for decentralised energy production), Siemens (small SOFC power plants), Vaillant, etc., and Viessmann (heating).

Germany is in an extremely weak position as regards the production of stacks apart from SFC. Stacks are only being produced as prototypes at individual university and non-university institutes. The majority of actors interviewed believed the absence of a German stack producer to be a major disadvantage. While some argue that it would be possible to produce stacks in Germany once a market seems ripe and the technology advanced enough, others see the major drawback of foreign stack production in the fact that technological knowledge at the system level is also situated abroad. Thus, the impulses for further development also come from abroad and, in some areas, German actors have to accept rather than define future technological paths.

Finally, one industrial actor group which is missing in the picture but which will become more important if the market enters mass production, are the providers of advanced *production technologies*. The traditional strength of the German machine tool industry and the results of the interviews indicate that the competence and production base in this area are regarded as being sufficient in Germany, or could be supplied on a sufficient scale if needed.

4.3.2.2 Public Research

Although there are strong individual university institutes in the area of fuel cells, the *university sector* as a whole is scattered and – with very few exceptions – focuses on individual technological aspects (such as certain materials etc.) rather than on the overall fuel cell system. There are only a few basic research actors concerned with the systemic nature of fuel cell technologies. The near market contribution to system development is thus in the hands of institutes and companies.

The application-oriented public institutes in Germany are – by definition – highly dependent on public and private research funding. As the majority of funding programmes finance co-operations that mostly include firms and institutes, the bulk of the work done in these institutes is – due to the commercial interests of the partners involved – directed at short- and medium-term applications, and thus towards improving existing solutions, rather than searching for radical innovations.

Given the need for immense *cost reductions* and *improvements in the long-term reliability* of the systems, there seem to be insufficient basic research actors for all the questions concerning the development of stacks or systems that require more "staying power".

However, in general, most companies consider the public research infrastructure in Germany to be sufficient, both with regard to the provision of basic knowledge and the development of near-market components. Only in a few cases such as in the area of stationary MCFC fuel cells did companies indicate that they have to rely entirely on in-house R&D due to a lack of co-operation partners in the more basic public research. On the other hand, especially in areas where the market has already taken off, such as PEM-based mobile applications, companies complain about research institutes not being application-oriented enough. This seems to indicate the traditional division of labour between private and public R&D rather than a systemic failure.

4.3.2.3 Interest Organisations and Other Corporate Actors

The fuel cell sector has developed a number of interest organisations and forums. German industry is organised in traditional business federations, the most significant of which are VDMA[31] or DECHEMA.[32] Industrial and research organisations come together in specific fuel cell organisations, such as the Brennstoffzellenbündnis (Fuel Cell Alliance) or one of the large number of regional initiatives.

---

[31] VDMA – Verein Deutscher Maschinenbauanstalten (German Engineering Federation).

[32] DECHEMA Gesellschaft für Chemische Technik und Biotechnologie e.V. (Society for Chemical Engineering and Biotechnology).

The members interviewed believe this Alliance is extremely important for being heard in the policy realm, and there seem to be ample possibilities to network and organise interest mediation and discourse (AMCG, 2002).

Recently, the former German Hydrogen Association reacted to the dynamics in fuel cell development and extended its activities more in that direction. A couple of years ago, the Federation changed its name to include "fuel cells". Here, the traditional interests of hydrogen actors are merged with the interests of many but by no means all of the fuel cell actors.

The overall structure, with VDMA and DECHEMA as the cornerstones and very heterogeneous memberships in fuel cell alliances, signals the horizontal character of the technology which combines various traditional strands of companies (machine tools, chemistry and electronics) and professions.

A new forum for the interaction of various stakeholders under the leadership of the German Ministry for Economy and Innovation is the "Strategy Council Hydrogen–Fuel Cells HYBERT" initiative that links the hydrogen and fuel cell strategic dialogues and is supposed to lead to a fuel cell vision and road map for Germany (see below).

At the European level, Fuel Cell Europe is by far the most important organisation for German actors, while the potential role of the European Hydrogen and Fuel Cell Technology Platform cannot yet be qualified. If the current proposal of the Framework Programme 7 (FP 7) is realised, the platform will be developed into a Joint Technology Initiative (JTI) on Hydrogen and Fuel Cells.[33] This initiative would mean a new funding structure at the European level to implement the research strategy defined in the Technology Platform. The JTI would go beyond the funding mechanisms of FP 7 and would mobilise a public/private partnership in the form of a joint venture between the Commission and European Industry to fund international activities. Organised extra-European discourse is rather weak and does not play a significant role at present beyond bilateral co-operations and partnerships.

4.3.3 Intermediaries, Financial and Other Institutions

As stated above and shown in the innovation system figure, there are no major traditional technology transfer organisations which could accelerate the technology cycle by transferring ideas from the research realm to the market. However, no interviewee indicated any lack of this kind of actor, since the technological status quo of fuel cells calls for direct co-operation between research and industry. As for financial institutions, the fuel cell community knows that many actors in Germany are rather risk aware, but this does not represent a major bottleneck because, after several phases of euphoria, the industry itself has become risk-averse in Germany, and banks simply reflect the caution they sense in industry and politics. It was mentioned in the interviews that the start of big research projects including

[33] See https://www.hfpeurope.org/uploads/835/JTI_QA_11JUL2005.pdf

SMEs was sometimes so late that the small companies were no longer present on the market.

Finally, the educational and training institutions in Germany supported in the ZIP programme have started to react to the challenges of a fuel cell economy. These challenges arise not so much in the area of training researchers – as this is still covered by specialisation within traditional disciplines – but in the area of maintenance and installation of fuel cells. Thus, for example, the Weiterbildungszentrum Ulm (Fuel Cell Education and Training Centre) or the Elektroausbildungszentrum Aalen offer training courses for tradesmen and engineers in order to adapt their skills. However, if the markets are to develop in the future, a greater public supply of such options is needed. One company which is successful in the market for mobile applications based on DMFC, for example, reports that most of its external co-operations are based on education and training.

4.3.3.1 Demand (Markets)

Demand and demand conditions constitute a major element of the innovation system. We have already described the developments and assessments with regard to market developments in Sect. 4.2.2 above.

4.3.3.2 Public Policy

The development of fuel cells in Germany would not be possible without public R&D funding. The markets are too distant, there are still many more basic R&D problems to be solved and the co-operation required is too costly and complex to be conducted in any meaningful way without public funding.

As can be seen in Fig. 4.16, fuel cells have been relevant in German research funding for more than 30 years according to the "Förderkatalog" ("Funding catalogue"). The steep rise in fuel cell funding since 2000 is due to the ZIP programme. PEMs have been promoted in projects since 1994, with a maximum in 1997 and with increasing sums up to the year 2004.

Because of the federal structure of Germany, not all projects are included in the Förderkatalog. According to Project Management Jülich (PTJ),[34] other funds for ongoing projects from 2004/2005 originate from the Federal states (84.2 million Euro), the German Federal government (12.3 million Euro) and the EU (34.8 million Euro). Of the three policy levels (EU, national and state level), the national level is by far the most important with several large programmes in place.[35] At this level, the policy institutions are scattered. However, there have been some recent efforts to better co-ordinate energy policy across the various ministries. For example, when

[34] http://www.nkj-ptj.de/H2/BZ-Projekte/

[35] Funds are also provided via the AIF (Arbeitsgemeinschaft industrieller Forschungsvereinigungen "Otto von Guericke" e.V.). According to information of the AiF, it is not possible to assign funds to specific technologies such as fuel cells because they are structured by research association rather than by technology.

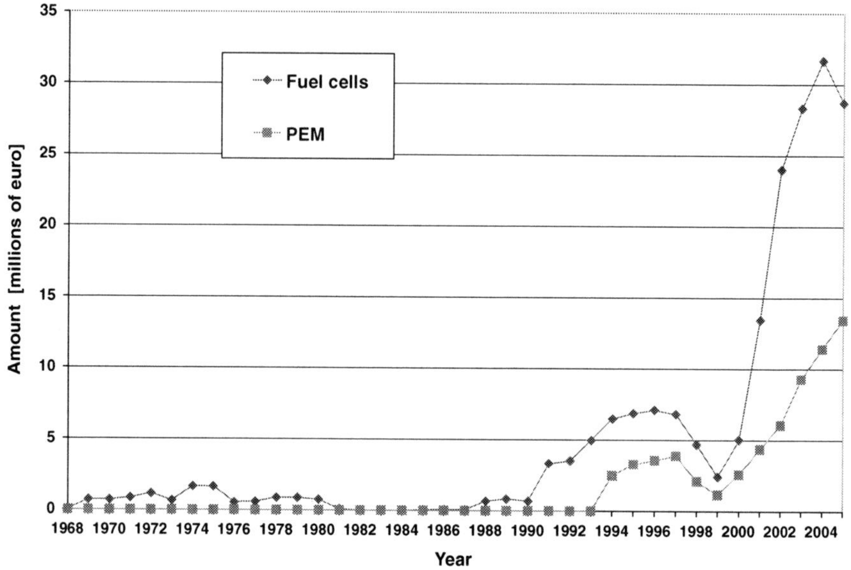

Fig. 4.16 Public funding of research projects on fuel cells and PEM according to the Förderkatalog, 1968–2004

financing individual projects, the relevant departments and ministries are contacted in order to check for double funding.

The most important actor – from the perspective of the interviewed actors as well as with regard to the budget dedicated to fuel cells – is the Federal Ministry of Economics and Technology (BMWi). The ZIP programme, which ran from 2001 to 2005, was the major source of funding for the fuel cell community in Germany during this period. There are overlaps in funding with the Federal Ministry of Transport with respect to mobile applications (mainly PEM technology). Furthermore, there is a division of labour with the Federal Ministry of Education and Research, which is responsible for basic research and for the non-university research organisations, while the BMWi concentrates its funds on application-oriented and other "promising" technologies. The major criterion for funding at the BMWi is that a project should result in a sound, reliable and relevant technology (as indicated by patents, prototypes etc.). The market potential in itself is relevant, but is not the major criterion since market developments cannot be foreseen by the Ministry ex ante. In the future, the Ministry is likely to be even more application- and product-oriented. For fuel cell technologies, the BMWi has not yet incorporated demand-oriented measures, but there are discussions and demands from industry for such measures in the future, resembling the instruments previously used for renewable energy technologies. This suggestion is discussed below.

Figure 4.17 gives an overview of the policy measures applied in Germany in the field of fuel cells based on a previous study (Wengel, 2006). The variety of instruments shown here is confirmed in interviews. Interviewees judge the mixture of the-

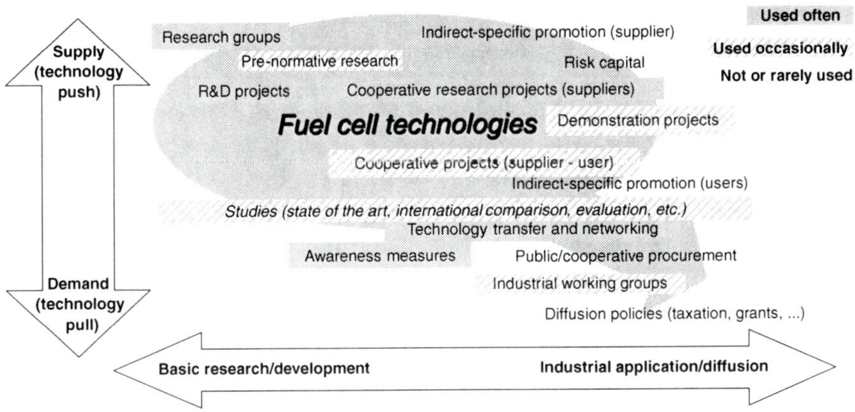

Fig. 4.17 Types of policy measures in Germany for fuel cell technologies
Source: Wengel, 2006

matic and horizontal programmes as flexible and appropriate. Furthermore, it shows that supply-oriented, R&D funding programmes – mainly funding co-operations – dominate and that demand-oriented activities are not overly important.

The policy mix reflects – by and large – the position of the fuel cell technology in the technology cycle, as it still concentrates on (co-operative) research, and thus the supply side. Some potential improvements of this co-operation will be dealt with below. The mix lacks, however, the long-term definition of potential demand developments and discursive developments of scenarios (with the Strategy Council Hydrogen–Fuel Cells as a promising approach here).

In the spring of 2006, some new initiatives emerged on the national level that might give the technology a new push. On April 1st, the Federal Minister of Transport, Building and Urban Development announced a new innovation programme for hydrogen and fuel cells.[36] The programme is a 10 year plan to boost early markets for hydrogen and fuel cells based on further R&D and market incentive programmes. This is in line with the recommendations of the "Impulse Circle Innovation Factor State" on lead markets within the initiative "Partner for Innovation". This initiative has brought together leading actors from ministries, science and industry. Their recommendations acknowledge that, overall, fuel cell technology needs a few more years to be technologically ready for the mass market, that measures need to be taken to close the gap as soon as possible and that, in parallel, lead market initiatives should be realised in selected areas via demonstration and lighthouse projects. This means that all the demand side measures put forward in these

[36] See press release of the ministry:
http://www.bmvbs.de/Presse/Pressemitteilungen-,1632.954868/Tiefensee-stellt-nationales-In.htm?global.back=/Presse/-%2c1632%2c2/Pressemitteilungen.htm%3flink%3dbmv_liste%26link.sKategorie%3d

recommendations will in fact be realised but it should not be forgotten that R&D efforts are still needed to bring the technology to maturity.[37]

The role the *federal states* play in Germany (i.e. the level of individual German states or Bundesländer) is very ambiguous. Some actors regard the initiatives – funding programmes and networking activities – at this level as very important and flexible, while others view the fragmentation and regional orientation of funding sources as counter-productive, since participation is dependent on the specific location within Germany which may exclude important co-operations. The capacities required for successful fuel cell development seem to be too distributed to be captured by state level programmes; improved coordination and opening up of these programmes is called for.

The International Dimension of Public Policy

International co-operations with public funding are mainly confined to the European level. Integrating international partners in national programmes is possible, but to a limited degree and mainly as subcontractors (suppliers). The US (DoE) is seen as being very open in this regard. But there is no consensus in the community concerning the opening up of national programmes to foreign actors. Some actors have reservations about the participation of foreign actors in national programmes, questioning the net effect for Germany. Others acknowledge the international division of labour and spread of expertise and are requesting easier access to national programmes based on evaluating each individual case against the contribution of a foreign actor to the overall objective of projects and programmes.

There is a broad consensus that the EU level should and will become more important in the future. For the German actors interviewed, it appears indispensable that their fuel cell activities are covered by the Framework Programme. This is not only because of the additional funding source, but also to ensure that the technological strengths and priorities of the German community are reflected in the Framework Programme. Interestingly, some actors regard having influence at the EU level as a means of shaping research and funding agendas in Germany as well, because they perceive the influence of the EU level on the priorities set in Germany.

However, the EU level has become less important and attractive for many actors over the last five years due to a high level of funding and lower financial burdens for consortium building and administration in German national programmes. The fuel cell community shares the view of many in assessing as excessive the administrative burden of the so-called New Instruments (Network of Excellence, Integrated Projects), which aim to build much larger consortiums than projects in previous Framework Programmes. Given the domestic orientation of national German programmes, there are indications that the public infrastructure and support for international co-operation is not sufficient. It remains to be seen whether a Joint Technology Initiative at the EU level (see above) could improve this situation.

[37] Working Group "Lead Markets" within the Impulse Circle "Innovation Factor State"(2006): Lead Markets Fuel Cells; manuscript.

4.3.4 Regulation

While norms and standards are not seen as bottlenecks, one regulation was mentioned as being problematic: the existing regulations for decentralised heating and power generation. According to the present law, tenants have to take heat but not the electricity from CHP systems.

At the same time, the possibilities for fostering innovation provided by regulations, for example in vehicle fleets, are not being exploited. Regulation has not been a major issue in the fuel cell community in the recent past, and processes such as norms and standards seem to operate rather well. One issue mentioned in the interviews was the difficulty small enterprises experienced when trying to work with international standardisation bodies: sometimes German standardisation takes place, but small companies are not able to finance ISO activities, and the national level did not seem worth the effort for them.

4.3.5 Co-Operations

Co-operation is a major characteristic of the fuel cell innovation system. This is especially true for research and development. Only one of the actors interviewed who is active in R&D reported relying solely on in-house capacities, and the majority of actors regard their own co-operation activities as more or less sufficient. There are two characteristic features. First, the *major* form of *research* co-operation is *between industry and research institutes*, rather than between companies, or between public research institutes. Of course there are some important and apparently successful co-operations between companies, but these are mostly very close to the market. For example, MTU and RWE Fuel Cells are planning to launch a joint venture for the development and production of MCFC for decentralised energy provision. RWE Fuel Cells is also co-operating intensively with IdaTech in the US on the integration of fuel cells into heating systems. PEMEAS, a provider of membranes for PEM and DMFC, also has various partners, two of which are from abroad (Motorola, Plug Power) and the DMFC application provider SFC has a strategic alliance with DuPont.

Strong co-operation between companies and public research institutes is to be expected for non-university, application-oriented institutes, but it is also found with university institutes, such as the ZBT Institute at the University of Duisburg, which was founded to work closely together with industry. However, it appears that the co-operation between public research institutes, both university and non-university, which would be needed for medium- to long-term research activities, is problematic to some degree due to the institutional backgrounds of the actors involved. Some of the major institutes have very little basic funding and need to raise money in the market and in application-oriented funded research projects.

Second, the vast majority of R&D co-operations are supported by *public funding*. This means that, without public funding, co-operation in the system would be severely limited. In addition, for the majority of the companies, publicly funded co-operations form a major part of their overall R&D activities. This is especially valid for component producers.

Furthermore, *international co-operations* do play a minor role, and if actors co-operate on an international level, then by far the most important funding source is the EU. Only those actors who have links to external partners via their company structure co-operate more intensively with external partners, such as – for example – RWE Fuel Cells (with IdaTech in the US), Siemens (with Siemens Westinghouse, Pittsburgh) and MTU (with FuelCell Energy Inc., Danbury). Interestingly, one company reports that they co-operate with foreign universities (UK), as these are more accessible and responsive to their needs.

One company which is already active and successful in a more mass oriented market (*SFC*) has only very limited research co-operations, i.e. its technology already serves a broader market need. This company has concentrated its co-operation activities on training installation and maintenance companies. There seems to be a knowledge gap on the part of the workmen who are responsible for the day-to-day maintenance of the new fuel cell based systems (Koschorke et al., 2005) and maintenance and installation may therefore represent a future bottleneck.

Co-operations between suppliers and producers at the various stages of the value chain are organised via the market, and the various forms of co-operation needed are realised within the system to a large extent.

4.3.6 Drivers of Future Markets

Based on the interviews, it seems that three actor groups are driving the future markets:

- first, the research organisations that not only assist industry in gradual improvements, but also offer new solutions for near market problems;
- second, the system providers in cars and in heating systems, i.e. both energy companies and heating system developers who see a future mass market for their applications;
- third, the producers of components who invest in improvements to make the overall systems more cost-effective.

While in the end development will be driven by the readiness of end users to invest in fuel cell energy technologies (see below), for this to happen, further investments are needed to improve costs and reliability.

4.4 Obstacles and Market Imperfections

In addition to the need for technology- and component-specific research described in the very first section of this paper, there is a further set of systemic and technological bottlenecks which affect the development of fuel cell innovations and markets. Not all, but some of these bottlenecks could be overcome with the help of public policy (see next chapter):

4.4.1 Risk-Aversion as a Result of Unfulfilled Promises

The anti-climax which occurred after the euphoria about the fuel cell economy seems to be the single most important challenge to the application of fuel cells in the future, because industry, end users and public funding programmes have since become much more risk-averse.

4.4.2 Framework Conditions (Energy Basis, Energy Price)

Unfavourable price developments, e.g. rising gas and falling electricity prices in the energy market, are another explanation for low demand. This relates to a further bottleneck, i.e. the energy basis for the fuel cells of the future. Renewable energies and hydrogen would have to be pushed for fuel cells to be environmentally advantageous. The role of hydrogen in fuel cell development has not yet been clarified: while the federation DWV is requesting that hydrogen be included in funding, other actors see more risks than benefits in this field.

4.4.3 Strong Developments in Alternative Technologies

Another source of uncertainty is the technological competition between technologies based on fuel cells and other alternatives. For many applications, such as portable micro cells, the market already offers technological alternatives that appear to be equally promising such as powerful batteries, and the abundant availability of electricity means that re-charging accumulators is unproblematic. As noted above, alternative systems in the field of decentralised combined heat and power such as gas motors, microgas turbines or Stirling motors are also not as expensive as fuel cells (by a factor of 5 or more, see Table 4.4).

4.4.4 System

As stated above, the majority of players interviewed agreed that the R&D capacities in Germany are sufficient, thus there is *no gap in research capacities* in principle.

However, there is a dilemma regarding the co-operation between public research and companies. On the one hand, some companies report that they would like to co-operate more intensively with institutes, but claim that public research institutes are not market oriented enough or too far behind in what they are doing. One company in the area of PEM applications claimed the market needs are more advanced than

Table 4.4 Examples of small CHP systems as an alternative to fuel cells, 2005

Type/Product/Manufacture	Investment Euro/kW_{el}	Efficiency [%] electric	thermal	total	World market	Remarks
Micro gas turbine Capstone μT 28-60/80 H	1,500	27	58	85	some 1,000	112 kW fuel provided. 30 kW electrical, 65 kW thermal
Stirling engine Solo V161	2,700	23	70	93	some 100	16–40 kW fuel provided. Variable 2–9 kW_{el} und 8–26 kW_{th}. Whisper Tech will produce 80,000 Stirling engines for the UK in the next years
Gas engine Senertec Dachs	2,800	27	61	88	some 10,000	20.5 kW fuel provided. 5.5 kW_{el}, 12.5 kW_{th}

Source: own results

the institute's capabilities and interests. On the other hand, one company in the area of MCFC (heating/power) complained of the lack of public research capabilities in its area which is more basic research oriented. Thus, while in the traditional area of non-university research – i.e. providing application oriented research with links to basic research – the capacities seem more or less sufficient in Germany, there does seem to be a gap at both ends of the research continuum – the very market oriented side on the one hand (e.g. DMFC applications, mainly on-board energy systems) and the basic research side on the other (MCFC power-heating applications) – at least in some of the fuel cell areas.

In most of the fuel cell variants, the German innovation system does not provide all the actors needed in the entire value chain. Above all, the absence of stack producers is seen as a problem in all fuel cell variants. In some areas, such as decentralised heating systems (Vaillant etc.), the number of producers is rather small and the market competition thus not dynamic or broad enough.

A minority of actors view the lack of venture capital for new ideas and small companies as a drawback. This is of course related to the general risk awareness and scepticism resulting from the delays in market creation.

4.5 Implications for Public Policy

Before the authors present some preliminary conclusions for energy-related R&D and diffusion policy, the next section briefly describes the major bottlenecks (see Sect. 4.4).

4.5.1 Bottlenecks and Opportunities

4.5.1.1 Technological Bottlenecks

There are two major issues with regard to the performance of the technology systems: *reliability and cost efficiency*. All three types of fuel cell systems still have to prove their long term technical *reliability and competitive cost* if they are to be produced on a commercial scale (see Table 4.5). This seems to have been underestimated by some stakeholders in the past. As long as the technical issues remain unresolved, operating costs and risks are too high and the necessary economies of scale from producing the components in highly automated production lines will not be achieved, thus delaying the potential market introduction. Keeping in mind that competitive technologies such as engine powered cogeneration, micro-turbines and Stirling engines are in the markets already, the fuel cell systems have to match at least the technical and cost performances of these technologies in their specific applications.

Table 4.5 Selected technical and economic bottlenecks in the market introduction of the three types of fuel cells, 2005

Type of bottleneck	PEM	MCFC	SOFC
– intensive utilisation and necessary recycling of platinum	yes	–	–
– cold start performance at low temperature	yes	–	–
– constant reliability of catalysts, fouling, corrosion, degradation	sulphur and oxygen content of fuel	corrosion resistance	materials for reduced working temperature
– overall complexity of the fuel cell systems	yes	yes	yes
– relatively high volume of the fuel cell compared with its competitors	yes	yes	yes
– availability of stacks mostly outside Europe	depending on start of large markets	depending on large industrial demand	depending on large industrial demand
– very high cost per kW compared to the competitive technology	yes, cost reduction needed: approx. 80–85%	yes, cost reduction needed: approx. 70%	yes, cost reduction needed: 60–70%

4.5.1.2 Lack of Demand Due to High Cost and Uncertain Operational Reliability

For many interviewees, mainly those from industry, one of the most significant bottlenecks is the *uncertainty as regards the readiness of users* to actually apply fuel cells in the future. End users still fear the high entry costs and have become cautious about promises made by industry. Furthermore, there is a *lack of awareness* concerning the ecological and long term economic advantages (after bringing down the cost curve by economies of scale) and thus no inclination to pay the higher price for fuel cells compared to other, presently available alternative technologies such as engine powered cogeneration, micro-turbines, or Stirling engines.

Thus, as stated above, the German market is not yet driven by end users with a strong demand for the energy-efficient technology – even though some component companies can sense a change currently taking place in this direction – and it lacks stack producers to push the development. The lack of demand within Germany is seen as one major reason for the absence of a stack producer in SOFC heating systems. One interviewee believed a demand of 500 installations here would be sufficient for a German stack producer to be established.

The various lead applications already on the market, such as on-board electricity for boats or camping vehicles, are still niche markets. Some of these niche markets, such as the fuel cell driven submarine or major CHP systems, are completely dependent on the state to purchase or sponsor demonstration projects. The signals from the demand side are still rather weak. Even if the technological and cost bottlenecks were overcome in the future, the widespread lack of confidence in fuel cell technology would still be a major obstacle to its market development.

4.5.1.3 Regulation

Most importantly, there are no "demanding" regulations which could help create a market for fuel cells in Germany. The Californian experience shows that tight *regulations* can spur market formation. The progress made in the vehicle application of fuel cells worldwide has been pushed by the rigorous Californian emission regulations due to the size and the lead function of this market. As these regulations are long-term by nature, they provide a clear future scenario for car producers and set strong incentives for an application strategy. However, as with other demand measures, any demand-oriented regulation would have to ascertain the readiness of the technology to be supported (see below).

4.5.1.4 Policy Bottlenecks

One of the key bottlenecks in Germany – as perceived by many different stakeholders – is the lack of a clear, explicit vision which could combine forces and lay out a roadmap for future progress. An example often cited was Japan, where the fuel cell community has both a clear framework and a roadmap (see below). The situation in Germany is problematic not least because of the scattered policy landscape, in which different ministries are active at the same policy level and the Bundesländer pursue their own individual programmes alongside Federal programmes. The co-ordination and/or lack of it, especially between Länder, is a major obstacle to a national strategy.

The public funding for fuel cells is seen as attractive by most of the actors interviewed, especially in the last few years, even though some actors, particularly those from industry, would have liked to see more money spent on fuel cell technologies. There are three major areas with room for improvement: *interaction between policy-makers* and *stakeholders,* activities to spur *public and private demand* for the future and *international co-ordination.* The actors involved in policy planning at the EU level regard the discursive policy planning activities of the EU as positive, but the lack of a comprehensive approach to defining a German position as negative.

Furthermore, a few actors claimed that concentrating funding on fuel cells while neglecting hydrogen research is problematic given the significance and relevance of hydrogen for fuel cell technology in the future.

Finally, path dependency represents a further challenge to public funding which is, of course, not confined to fuel cell technologies. For example, a lot of investments have been made in low temperature technology, see Fig. 4.14. Since then, however, many actors have come to regard the high temperature technology as more promising, but a change of direction would be very costly because many partners are involved and turning industry in a new direction would cause major sunk costs and frustration in some parts of the industry.

4.5.1.5 International Co-Ordination

Importing technology has proven problematic since the market conditions and applications are different in each country and adjustments tend to be expensive. At the same time, further developments are costly or even impossible without international co-operation. Thus, the major bottlenecks here are a lack of co-ordination of national programmes and insufficient funding possibilities for

- extra-European co-operations and
- small-scale projects with one or two international partners.

4.5.2 Policy Recommendations

Not all of the bottlenecks mentioned above can be remedied by public policy action. However, they do show possibilities for policy to focus and re-adjust. As a basis for discussion, the following – still preliminary – set of policy recommendations is proposed:

(1) *Definition of a fuel cell vision and strategy*

Following international examples where a vision and a roadmap for fuel cells has been designed and coordinated with stakeholders and between ministries,[38] the discourse in Germany should be intensified which aims at a *long term vision* and *medium-term strategies and roadmaps*. This vision should be feasible and realistic and include markers which, when achieved, automatically result in questioning the progress and instigating revisions. Such revisions should also take a look at the achieved state of the art of competing technologies. This would not only help to

---

[38] For example, in Japan, the METI chaired a Fuel Cell Commercialisation Study Group (28 members from industry, universities and other public institutes) and the Fuel Cell Commercialisation Conference of Japan (134 institutional and individual members) resulting in a roadmap including public procurement as a market creation instrument until 2010. The US National Hydrogen Energy Roadmap from November 2002 and the implementation plan "Hydrogen Posture Plan" of the Department of Energy are based on a national "Vision and Roadmapping" process. This plan has defined several phases and concrete targets, including 100 specific technological targets and concrete commercialisation plans up to 2015. Entry into the market is supposed to begin in 2010. Both in Canada (Hydrogen und Fuel Cell Coordinating Committee) and the UK (High Level Steering Group), high level expert forums are working towards similar strategies, including start up financing for the participation of industrial federations. In all these countries, co-ordination mechanisms between ministries are in place. For further details, see Wengel (2006).

better define research priorities, but would help to instil stakeholders with a greater sense of confidence again.[39]

The HYBERT approach promises the discourse necessary for such a vision.[40] It is a step in the right direction and as such is welcomed by all the stakeholders. HYBERT, which is now known as "Strategierat Wasserstoff-Brennstoffzelle" (Strategy Council Hydrogen–Fuel Cells), should co-ordinate and harmonise the different activities.[41] It has started to do so by developing a National Development Plan H2 and FC Technologies in July 2006.[42] The analysis has shown that such a process is challenging, and needs to be visible and transparent to be regarded as legitimate. Furthermore, it must include experts beyond the circle of interested stakeholders (*think tank culture*) and it should set up interfaces for international co-ordination.

(2) *Future actions to spur public and private demand – sound preparation and appropriate timing*

Market development also represents a major future challenge. Even after basic technological problems have been solved, market entry is still very costly due to the lack of economies of scale. Public action to spur demand, however, would have to take place in the right window of opportunity. As the ultimate objective is to trigger a self-supporting market, such a demand programme must not be premature. This is also acknowledged by the recommendations within the impulse circle "innovation factor state" (Working Group Fuel Cells 2006), which calls for market measures, but sees the necessity for further R&D efforts before pushing for large scale markets. Furthermore, demand articulation processes – for example in the form of an interactive, constructive technology assessment (Sundermann, 1999; Smits, 2002) including *all* kinds of stakeholders – could be organised in order to construct a realistic picture of future demand and to make all parties involved aware of expectations and scenarios. The Strategy Council Hydrogen–Fuel Cells initiative would be a good nucleus for such a broader approach to integrating demand.

Moreover, it appears that markets start to grow once a technology is able – technically speaking – to meet its demand at an affordable price. From the indicator analysis (growing number of publications and patents etc.) and given the time horizon for potential market applications mentioned in the interviews (2010–2015 for most variants and applications), it seems that a number of technical problems still remain unsolved. Thus, demand orientation should not be too broad, but geared towards specific technologies that require scale effects rather than major technological improvements.

---

39 Already the coalition treaty of the new government stipulates that joint initiatives of science, industry and politics shall define strategies to keep or increase German technological and market leadership in new technologies (http://www.bundesrat.de/Site/Inhalt/DE/1_20Aktuelles/1.1_20Foederalismusreform/NI/Koalitionsvertrag,property=Dokument.pdf).

40 Other associations such as the Fuel Cell Alliance are important because they help to structure fuel cell activities in Germany.

41 http://www.nkj-ptj.de/datapool/page/7/H2_BZKonzept8Mai2006.pdf

42 National Development Plan H2 and FC Technologies Version 1.1 of the Strategy Council 25 July 2006. http://www.nkj-ptj.de/index.php?index=76

A potential demand-oriented programme would, of course, have to be one cornerstone of the interactive strategy development mentioned above, and industry would have to provide evidence of the long-term benefits of their systems, e.g. for decentralised power-heating systems, and of the resulting economies of scale.

Based on the results of the indicator analysis and the interviews, the most suitable areas for such demand-oriented programmes are those that promise a mass market in the future. A preliminary list of technologies includes:

- low power applications or applications in which costs are not the major obstacle,
- decentralised CHP systems based on PEM (MCFC, SOFC) and
- mobile applications.

(3) *Research Funding Programmes*

In general, research funding should focus on the *technological bottlenecks* mentioned above, i.e. it should *prioritise projects* that lead to *cost reductions and greater reliability*. This would mean higher priority being given to explicit targets in projects rather than inputs. Furthermore, the analysis has clearly shown that the approach of *focusing on funding co-operations* is successful and should be continued. In addition, funding schemes should be tested in which application-oriented research institutes could be included in more basic research activities, thus *fostering the interaction within the public research community*.

There are calls for *stronger links between research funding* and *application oriented activities*. Even when markets are in the process of being formed, scale effects alone may not result in the breakthroughs needed for costs and reliability in many instances, since these problems often only emerge during the application phase. At the same time, however, funding decisions even at the more basic research end should take market potentials and realistic market expectations into account for certain technologies. Thus, *the division between basic research and market application should be dismantled as far as possible*. This means that basic research should be closer to the concrete problems arising from application wherever possible and that basic research results should be fed into the market introduction of technologies.[43]

Finally, coordination should be improved not only between Federal ministries, but also between ministries at the level of the individual Bundesländer, as many actors report that other states' programmes are not accessible to companies/institutes outside the specific region concerned, even though the company or institute could contribute to and benefit from the programme.

(4) *International co-ordination*

German policy should aim to co-ordinate a distinctly German position on fuel cell priorities in the Framework Programme more intensively and transparently. Second, schemes should be developed that enable extra-European co-operation and

[43] This may be difficult to handle as experience has shown that former networks concentrating on basic research but inviting industrial participation did not sufficiently meet the interests of industry, just as calls that invited co-operation between industry and basic science were not attractive enough for basic science.

small-scale international co-operation with multi-lateral national funding. Fostering ERA-NETS may be a major step in this direction. Furthermore, the efforts of the European Technology Platforms in the area of fuel cells should be closely monitored, as they bring together a broad variety of factors with an international perspective, including potential user groups.

(5) *Regulation*

As demand measures, regulations fostering fuel cells must take into account the state of the art of the technology in order to avoid over-burdening producers and purchasers. Two potential lines of action are conceivable in the future: first, national policy could influence European regulation to set stricter standards for future vehicle fleets that demand alternative technologies such as fuel cells. Second, policymakers should examine whether current national regulations are suited to a complex technology system like the fuel cell. For example, the current regulations affecting household heating oblige tenants to take heat but not electricity from co generation systems in their building.

References

AMCG. (2002). *Markteinführung von Brennstoffzellen-Produkten: Auswirkungen auf den Maschinen- und Anlagenbau*. Voruntersuchung.

Beer, M., & Mauch, W. (2006). *Ganzheitliche und dynamische Bewertung neuer Energietechnologien der KWK mit Brennstoffzellentechnologie*. FfE, München.

Brennstoffzellen bündnis Deutschland: Brennstoffzellen-Bündnis Deutschland ruft Politik auf, die Strategie zur Markteinführung umzusetzen. Presseinformation vom 14.12.2004.

Frost & Sullivan. (2001). Stationary and portable Fuel Cells – Developments, Markets and Opportunities. Report D226.

Gummert, G., & Suttor, W. (2006). *Stationäre Brennstoffzellen – Technik und Markt*. Heidelberg: C.F. Müller Verlag.

Jochem, E., Landwehr, M., Tönsing, E., Nöltscher, Ch., & Vollmar, E. (1997). Energy Scenarios with reduced CO2 emissions until 2050. (in German) *Energiewirtschaftliche Tagesfragen, 8*, 474–480.

Koschorke, W., Bünger, U., Marscheider-Weidemann, F., Pirk, W., Roser, A., & Zerta, M. (2005). *Anforderungen an das Handwerk durch die Innovation Brennstoffzelle*. Stuttgart: IRB-Verlag.

Krewitt, W., Nitsch, J., Fischedick, M., Pehnt, M., & Temming, H. (2006). Market perspectives of stationary fuel cells in a sustainable energy supply system – long-term scenarios for Germany. *Energy Policy, 34*, 793–803.

Krewitt, W., Pehnt, M., Fischedick, M., & Temming, H. V. (Hrsg.). (2004). *Brennstoffzellen in der Kraft-Wärme-Kopplung – Ökobilanzen, Szenarien, Marktpotenziale*. Berlin: Erich Schmidt Verlag, Berlin.

Riensche, E. (2006). Ganzheitliche und dynamische Bewertung neuer Energietechnologien der KWK mit Brennstoffzellentechnologie. Technische Herausfordungen bei der Entwicklung der Brennstoffzellen bis zur Marktreife. FZJ – IWV 3. Jülich

Smits, R. (2002). 'Innovation studies in the 21st century. Questions from a users perspective'. *Technological Forecasting and Social Change, 69*(9), 861–883.

Sundermann, C. (1999). Constructive Technology Assessment, in: Stephan Bröchler, Georg Simonis, Karsten Sundermann (Hg.): Handbuch Technikfolgenabschätzung, Berlin 1999, edition sigma, S. 119–128.

Temming, H. V. (2005). *Analysen und Simulationen zur Bewertung der Einsatzmöglichkeiten und Marktpotenziale von Brennstoffzellen in der Industrie*. Bochum: LEE Uni Bochum.

Wengel, J., & Schirrmeister, E. (Ed.). (2000). *Innovationsprozess vom Verbrennungsmotor zur Brennstoffzelle. Chancen und Risiken für die baden-württembergische Industrie.* Karlsruhe: Fraunhofer ISI.

Wengel, J. (2006). Das Beispiel Brennstoffzellentechnologie (The Example of Fuel Cell Technologies). In J. Edler (Ed.), *Nachfrageorientierte Innovationspolitik. Büro für Technikfolgen-Abschätzung beim Deutschen Bundestag (TAB)*, Berlin.

Wurster, R. (1999). PEM fuel cells in stationary and mobile applications – Pathways to Commercialization. Sixth International Technical Congress – BIEL'99 – 13th – 19th September 1999 -Bienal de la Industria Eléctrica y Luminotécnica. CADIEM Camara Argentina de Industrias Electromecánicas. Biel.

Chapter 5
Passive Houses and Buildings

A market survey by Fraunhofer ISE (2004) showed that the number of Passive and low-energy buildings will increase significantly in the future. For this study 180 architects, contractors, prefabricated housing manufacturers, designer engineers and manufacturers of ventilating units were interviewed together with 25 consultants, scientists and representatives of funding institutions. The experts forecast that around every fifth new building will be a Passive house, and every third will be a lowest-energy house by 2010. The following advantages to investing in this technology are listed: low heat requirements, increased living comfort, low additional costs and reduced dependence on increasing energy prices/costs.

Based on this survey, scenarios for the market diffusion of Passive and lowest-energy houses were developed by Büro für Solarmarketing (Freiburg). The worst case scenario result for the number of Passive houses/year is an increase from 1,300 today to 60,000 units by 2010. The number of new 3-litre houses/year is expected to grow from 3,500 to 100,000 in 2010, i.e., half of new buildings will be Passive houses or lowest-energy houses if their potential can be successfully realised by offensive marketing (Bühring, Leuchtner, Krug, & Schüle, 2004).

5.1 Description of the Passive House Technology and Energy Economic Relevance

Approximately one quarter of the primary energy consumption in Germany is use to heat buildings. Almost 90% of this could be saved in the long term using modern technologies and concepts. The passive house is one important option here. The energy-related refurbishment of older buildings plays a bigger role because of the relatively low rate of new building developments compared with the large stock of existing buildings. Even if this chapter is mainly concerned with passive houses as new buildings, all the components and processes are also used in building refurbishment.

A passive house is a building which is constructed so as to achieve a comfortable interior climate *without a separate* active heating device. It has an annual space heat requirement of 15 kWh/(m^2a) or less. Including the preparation of hot water and all

E. Jochem (ed.), *Improving the Efficiency of R&D and the Market Diffusion of Energy Technologies*, DOI 10.1007/978-3-7908-2154-3_5, © Physica-Verlag Heidelberg 2009

household appliances, the passive house has an overall primary energy requirement of 120 kWh/(m^2a). A passive house achieves the low energy requirements via the kind of components installed in the building and the building itself as an integration of the components guided by two principles: first, minimize heat losses, second, maximize heat gains. Three components prove crucial for the passive house. (1) Superinsulation and excellent air tightness, (2) heat recovery systems and (3) passive solar gains in winter. If, in addition to this, highly efficient household appliances are used and the remaining energy requirements are met using renewable energy resources, the passive house is even more energy-efficient and environmentally friendly.

The realisation of passive houses or buildings places high demands on the components used (http://www.passivhaus-info.de/):

1. Thermal insulation: U-values under 0.1 W/(m^2K) as a rule
2. Construction without thermal bridges to a large extent
3. Proven excellent air tightness using pressure tests: pressure test air exchange at 50 Pa should show a pressure difference smaller than 0.6 h^{-1} better 0.3 h^{-1}
4. Glazing with U_g values below 0.8 W/(m^2K) at high total solar energy transmittance (g> 50%)
5. Windows with U_w values below 0.8 W/(m^2K), i.e. insulated window frames installed without thermal bridges
6. Highly-efficient heat recovery from ventilation system ($\geq$75%) at low electricity consumption ($\leq$0.45 W/m^3)
7. Lowest heat losses when producing and distributing domestic hot water: insulated hot water pipes, well insulated hot water storage tank
8. Highly-efficient use of household electricity: where possible, use of energy-saving electrical appliances and lamps

When considering the additional monthly expenses for a passive house in the form of higher loan interest rates and repayments minus the financing available at the moment under sponsorship programmes and energy cost savings, a passive house can still be profitable despite its high additional cost since the following points also have to be taken into account (see also Fig. 5.1):

- avoided costs for installing a chimney
- avoided costs for gas connection or fuel oil storage
- no reinvestment in a heating system for approx. 20 years
- no costs for cleaning the chimney
- no maintenance costs for the heating system.

In this way, the additional costs for a passive house can be covered by the cost savings, particularly by the avoided energy costs.

Due to the thicker walls owing to increased thermal insulation, in the passive houses constructed at the moment, either the living area will be smaller or the building plot will have to be correspondingly larger, which can lead to an additional financial burden especially at high real estate costs. This can be resolved in the future using high performance insulation.

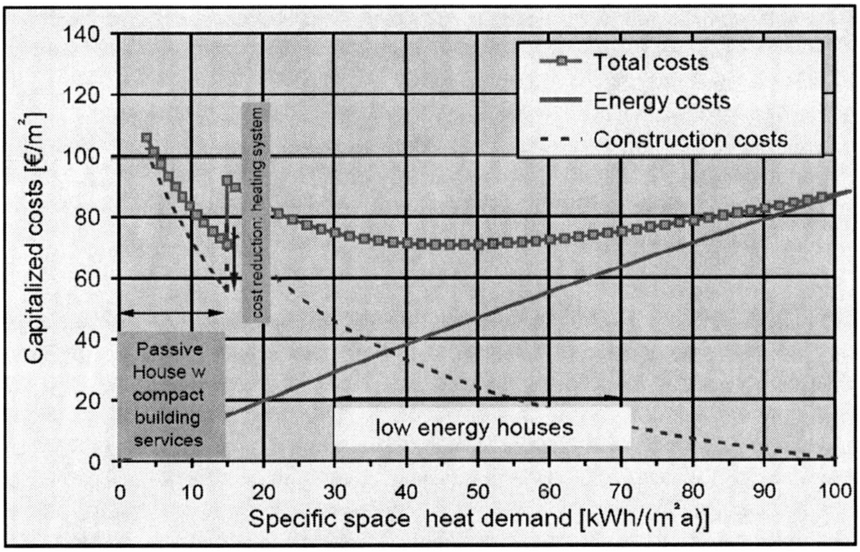

Fig. 5.1 Typical dependence of total, capitalized costs on the heat demand of a terraced house (passive house case with compact building services)
Source: Schnieders/Hermelink: 2006.

Compared with a conventional house, the additional components of a passive house, especially thermal insulation materials and the greater mass due to energy storage requirements have a higher energy demand during production. This is more than compensated for by the avoided heating system as well as by the saved heating energy. See Siraki, Wagner (2006) for more details.

In Germany, around 35% of the final energy demand is required to supply space heat. Passive houses can save 90% of the average heating energy demand of 150–200 kWh/m^2 and year of the existing building stock. Due to the low house building rate in Germany, this can only take place in the very long term; quicker success is possible by modernizing the heating technology in the existing building stock. Even if the passive house standard seems difficult to achieve in a lot of cases – 25 kWh/m^{2}a seems to be a realistic value for refurbishment – many passive house components and methods can still be transferred to the existing building stock. The energy savings achieved in existing demonstration projects of retrofitting older buildings are in the range of 75–90%.

5.1.1 Superinsulation

There are a multitude of outside wall constructions which are suitable for low energy houses and passive houses:

- Outside wall thermally insulated combined system (external thermal insulation compound system with 30 cm insulation).

- Lost formworks made of EPS foam, which are filled with concrete at the building site. It is not particularly difficult to reinforce the EPS by several centimetres in order to meet the passive house standard.
- Prefabricated multi-layer wall elements with polyurethane insulation.
- Wall elements with timber studs, e.g. double-T-beams and more than 30 cm thermal insulation.
- High-tech version: *vacuum superinsulation*, which is already able to reach the necessary low heat transition coefficients at thicknesses of 2.5–4 cm.

Highly efficient evacuated thermal insulation can be used in the building sector in the form of flat vacuum insulation panels (VIP). By using such panels, the thickness of insulation can be reduced by up to a factor of 8–10 compared with conventional insulating materials, e.g. from 30 cm to 3–4 cm. This can be done by using microporous core material, usually SiO_2, which is contained in an evacuated and airtight envelope. The challenge for research consists in guaranteeing the ultra low thermal conductivity for 20 plus years. The VIPs can be integrated into the building envelope in various ways. Within the IEA Annex 39, VIP measures are realised in a large number of buildings. There are also some excellent examples for the use of VIPs in refurbishment, as shown by the two prizes awarded for building physics in 2005.

5.1.2 Heat Recovery Systems with Active Heating or Cooling Components

It is essential to have a ventilation system with heat recovery in a passive house. A small downstream heater in the supply air is sufficient to heat the entire building. The heat for the **downstream heating of supply air** may for example come from the hot water supply system or can be produced by special small power units as follows:

One possibility is to use a heat pump: a **mini heat pump** (approx. 1,000 W condenser, a compressor from a cooling unit series) removes source heat from the exhaust air of the air/air heat exchanger. This is warmer than the outside air and contains the total latent heat of the steam emitted inside the house. If this exhaust air is cooled down from 20 to 2°C, a heat flow of 500 to 800 W can be extracted at the evaporator. With such a system it is possible to meet the total demand for hot water and the remaining demand for space heat at an electricity consumption of 1,000–2,200 kWh/a. The appropriate appliances are currently being produced by several medium-sized enterprises.

One alternative is **direct heating with natural gas in a compact unit** – corresponding highly efficient appliances are still being developed. It is also possible to supply heat using conventional gas or oil condensing boilers or district heat.

Another alternative is a **mini gas heater** used for downstream heating of the supply air. The simplest case is a slightly adapted auxiliary heater such as those used in lorries or caravans.

The application of an air conditioning system may be necessary due to the rising summer temperatures and increasing standards of quality and comfort, but also to correct errors made by the residents in shading or ventilating rooms, even if this should actually be avoided for reasons of energy efficiency by using passive structural measures and automatic shading devices. One possibility is the use of the ventilation systems for **sorptive cooling**: the outside air releases some of its humidity to the absorbers in the **sorption wheel**. As a result the temperature of the supply air rises. A thermal wheel which has not been coated with absorbent cools the supply air. The required cooling potential is extracted from the exhaust air which is piped through the wheel to the other side. This was cooled beforehand in a humidifier. The outside air which has been dehumidified in the sorption wheel and cooled in the thermal wheel is then humidified to the desired level which cools it down again. The exhaust air is heated in a heat exchanger by solar energy or by a gas burner in order to regenerate the sorption wheel and is blown outside loaded with steam. Depending on the outside temperature, 12–19°C supply air is possible with 100% fresh air. At a load of approx. 75%, the sorptive air conditioning system has a coefficient of performance of 1. In winter, the thermal wheels (without humidification) operate to recover heat and humidity.

5.1.3 Windows and Passive Solar Gains

Modern **triple glazed windows** achieve U values of 0.55–0.8 W/(m^{2}K). These windows have two low-e layers and are filled with argon or krypton. The thermal weak points are the spacers of the glazing which are made of aluminium for reasons of air tightness but which is also a good thermal conductor. Last, but not least, so much progress has been made in the research on vacuum glazing that within the next couple of years there will be windows available with U values of 0.4 W/(m^{2}K).

Particularly well insulated window frames were developed which also restrict the glass edge seal losses by increasing the depth to which the glazing is inserted into the frame. The losses at the edge seal are decreased still further by thermally separated spacers (so-called "warm edge system").

Work is being done at present on special coatings of panes which are automatically controlled by light or temperature, or which, via an electrical current, can greatly alter their reflection capacity to incident sunlight; these coatings serve the maximum exploitation of solar power, but at the same time also serve as solar protection to prevent overheating, a problem which will become increasingly significant.

5.1.4 Competing Concepts

There is a seamless transition of houses according to the provisions of the Energie-Einspar-Verordnung (Energy Conservation Ordinance (EnEV)), the Low Energy Standard (45% below the EnEV values), through the Energy-Saving Houses

Standard KfW60 and KfW40, 3 litre houses, passive houses, zero heating energy houses up to "plus energy houses". Solar houses cut across this classification. The philosophy behind the individual concepts may differ. The EnEV only sets the primary energy demand for heating and hot water at certain minimum insulation values of the building's envelope. This can be achieved either by e.g. sophisticated measures carried out on the building shell, or, at a lower construction effort, by using renewable energy sources such as e.g. wood pellet furnaces. The KfW houses are oriented towards these possibilities but stipulate a maximum primary energy demand of 40 or 60 kWh/m^2 and year and more demanding minimum insulation values of the outer shell, which have to be 30% better than the EnEV for the KfW60 house and 45% better in the KfW40 house which meets the passive house standard.

In passive houses, the heating demand has been lowered to such an extent as a result of constructional measures that *no separate heating system* is necessary. This is possible from around 15 kWh/m^2 and year which corresponds to a heating load of around 10 W/m^2 under prevailing climatic conditions, since then the internal thermal loads of the inhabitants and household appliances as well as the downstream heating of the supply air in the ventilation system can cover the heat losses in winter.

Other concepts consider a larger or smaller share of solar energy, and sometimes also other renewable energy sources. The idea here is the realisation of minimum costs at a given maximum heating or primary energy demand or minimum greenhouse gas emissions without tapping the full potentials of constructional measures. Perhaps these other concepts anticipate the difficulty many people have in imagining a house without an active heating system. Since, however, even the use of biomass is not completely free of environmental impacts and furthermore is also ultimately limited and the use of photovoltaic systems is also possible in a passive house, the passive house approach actually often represents the most sustainable option. As long as there are doubts about the cost efficiency or the "error friendliness" of passive houses and buildings, other concepts – like low energy houses and buildings – should continue to be pursued in order to provide alternatives to houses under the EnEV (see Exhibit 5.1).

Exhibit 5.1 Alternatives to passive houses or buildings

Passive houses, low energy houses and conventional buildings can be grouped according to the housing services they deliver and the energy efficiency they achieve. Pricing information is not contained in this sketch. It is clear, however, that of the three types, passive houses excel in both housing services and energy efficiency. The energy efficiency is achieved by the required technological characteristics. The ventilation system and the increase in comfort put the passive house or building a notch ahead of low energy houses.

Despite this, low energy houses are still the major competing technology for the passive house.

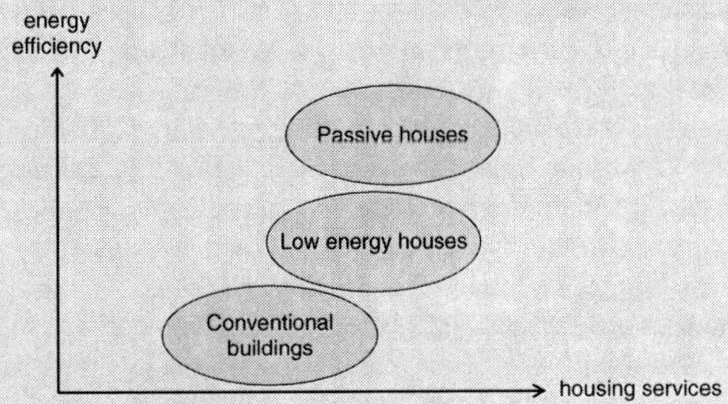

Housing services and energy efficiency are indicated purely to serve as a visual aid. The relative position of the three technologies is roughly in accordance with reality; the distances between the technologies cannot be detected from the above figure.

5.2 Analysis at the Micro Level

5.2.1 Technological and Economical Challenges

Before explicitly discussing the technology cycle and the technological innovation system of passive houses and passive buildings, we have to discuss the nature of the passive house as an innovation, since this will facilitate the further discussion below. It will structure our approach towards the assessment of the technological cycle phase. It will also help to highlight the peculiarities of the innovation system. Also, as will be elaborated below, the type of innovation has a bearing on the structure of the innovation system.

As shown in the previous section on the technical description of passive houses, passive buildings as a product are a system consisting of a number of components, where the components are physically distinct parts of the system such as walls, insulation, air conditioning, windows and doors etc. Based on the physical distinction between the system as an integration of the components and the components themselves, we can also distinguish two broad types of knowledge relevant for the production of a passive house or building: (1) **component knowledge** and (2) **systems integration knowledge** which refers to knowledge about the integration of the components into a system.

Passive houses are an **architectural innovation** in the sense of Henderson and Clark (1990). Although confusion may arise in the context of construction and building innovations, the term *architectural innovation* does NOT refer to the fact that passive houses are designed and planned as a product of architects' activities. The term *architectural* is used by Henderson and Clark to distinguish a particular type of innovation from incremental, radical or modular innovations. The particular feature of architectural innovation is that the component knowledge remains more or less untouched by the innovation. The source for the novelty is the knowledge of how to integrate the components into a product. The key concept in architectural innovations is that existing components are linked and configured in an essentially new way. Further below in the discussion we will use the term *systems integration innovation* instead of *architectural innovation* in the sense of Henderson and Clark.

In the case of a passive house, the new configuration of existing components is the consistent implementation of energy saving components in a planned and coordinated way so as to minimize the heat loss through the building envelope and optimize passive heat gains from solar irradiation and household appliances. As such, passive houses are a refinement and further development of low energy houses (Schnieders & Hermelink, 2006). As part of the systems integration, the novelties in passive buildings include an airtight building envelope, the avoidance of thermal losses through thermal bridges and superinsulation, thermal insulation of windows and doors and the reduction of the building surface relative to the building volume. It also comprises the planned and optimized implementation of ventilation installations with highly efficient heat recovery. Although the term *systems integration innovation* draws our attention to the changes in the linkages of the components and their configuration rather than to the core design concepts of the components themselves, the changed links between the components may actually trigger the evolution of some of the components such as efficiency increases in heat pumps and ventilation appliances, the development of new types of insulation materials and improving the existing two-layer glazing of windows. The systems integration innovation nature of the passive house is emphasized by the fact that "the simple combination of passive house certified components does not necessarily result in a passive house" (Horn, 2005, p. 403).

The change in the configuration of the components also can be triggered by developments in one of the components. As the passive house is a refinement of the low energy building, the trigger for the development and production of passive houses can be attributed to the consistent realisation of ideas and concepts already being implemented in low energy buildings.

5.2.2 Passive House Technologies in the Technology Cycle

The indicator-based assessment of the technology life cycle below will reflect the previous discussion about the systems integration nature of the passive house innovation. As discussed above, the primary novelty about a passive building is the systems integration knowledge required to combine and link the individual components needed to achieve the relevant passive house properties. Changing the architecture of a product to yield improved performance tends to require well-developed components. In terms of the position in the technology cycle, the components are likely to be in a fairly late phase of their technological development, notwithstanding the fact that the new overall design and the new linkages of the components may induce further potentially incremental changes to the individual components.

By and large, the passive house relies on a number of well-developed components. Both in the patent analysis and in the publication analysis, custom search strategies had to be developed to match the nature of the passive house as a systems integration innovation.

5.2.2.1 Patent Dynamics

In line with the insights gained, the main aim of analysing patent dynamics is to identify the technology cycle phase of each individual component.

The search strategy identified EPO patents based on the technology sub-classes for wall systems, glazing, ventilation systems and doors & windows. Exhibit 5.2 shows the technology classes and the components covered. The development of patent applications for wall systems, ventilation and air conditioning components, glazing, as well as doors and windows shows a steady and pronounced growth between 1979 and the end of the 1990s before some saturation seems to occur at the level between 40 (ventilation systems) and 120 patents per year (windows and doors, see Fig. 5.2).

Exhibit 5.2 IPC classes for patent search of the components of passive houses

Walls E04B-002/96/IC; F24F-005/00/IC; E04B-001/76/IC; C09D-005/32/IC; B28B-019/00/IC; F24J-002/04/IC; E04B-001/74/IC;

Glazing C03C-017/36/IC; E06B-003/66/IC; G02B-001/10/IC;

Ventilation & AC F24F-012/00/IC; F24F-011/00/IC; F24F-013/08/IC;

Doors & windows E06B-003/263/IC; E06B-003/54/IC; E04B-002/88/IC; E06B-003/82/IC; E06B-001/30/IC; E06B-001/62 /IC; E06B-003/56 /IC; E06B-009/264/IC; B29C-044/38 /IC; E06B-003/22 /IC; E06B-003/964/IC; E06B-003/30 /IC; E04B-001/80/IC; E06B-003/968/IC; E06B-003/273/IC; E06B-001/02 /IC; E06B-007/23 /IC;

As the development of passive houses builds on given and fairly well developed components, it is important not to try and identify patents immediately related to passive buildings. Instead, it is crucial to identify the development trends in the technology fields which eventually end up being used in passive buildings. The patent search conducted thus tracked the development of identified technology classes (IPC) which were defined independently.

Over the past 25 years, a dynamic development can be observed for each of the components shown in Fig. 5.2. The substantial increase in patenting in the 1990s is followed by a subsequent drop which indicates that the passive house components are based on sound and well elaborated technologies. Any subsequent modifications of the components induced by their use in passive buildings cannot be tracked using patent analysis.

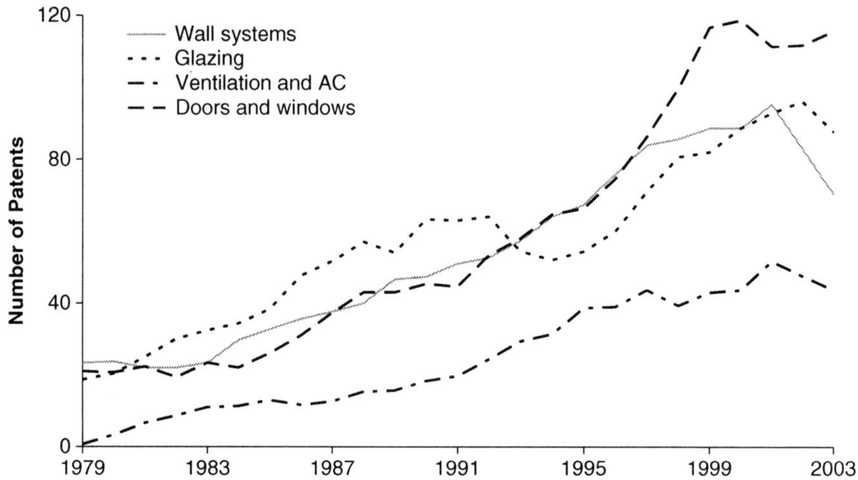

Fig. 5.2 Patents for components to be integrated in a passive building, 1979–2003

From the mid 1980s to the beginning of this millennium, wall system related patents and glazing related patents have an annual average growth rate of 6.0% and 6.4% respectively. Ventilation systems and doors & windows experienced a higher annual growth rate of around 9.5% and 11.8% respectively. The growth rate of the latter exceeds the patent application growth rate at the EPO for all technologies, whereas wall systems and glazing are within the overall trend.

5.2.2.2 Scientific Publications

The identification of scientific publications related to passive houses also required a custom search strategy. The search strategy employed to identify passive house publications is shown below:

Figure 5.3 illustrates the development of the number of publications in the domain of passive houses from some 10 per year in 1979 to about 60 per year around the year 2000. It reflects the number of retrieved documents after the iteration of step 3. As such it represents the broader search approach.

The more focussed search shows the development[44] in the number of scientific publications with a direct focus on passive houses and their technologies to have experienced its first peak around the early 1980s. There is a slight reduction at the end of the 1980s and in the early 1990s and there has been a continuous increase since the mid 1990s. This is a subset of the publications drawn from the set of publications shown in Fig. 5.3. The direct link to passive houses in these publications was established by processing the information contained in the title and abstract of the papers as described in step 4 (1) above. The peak of passive house related pub-

[44]Not depicted here.

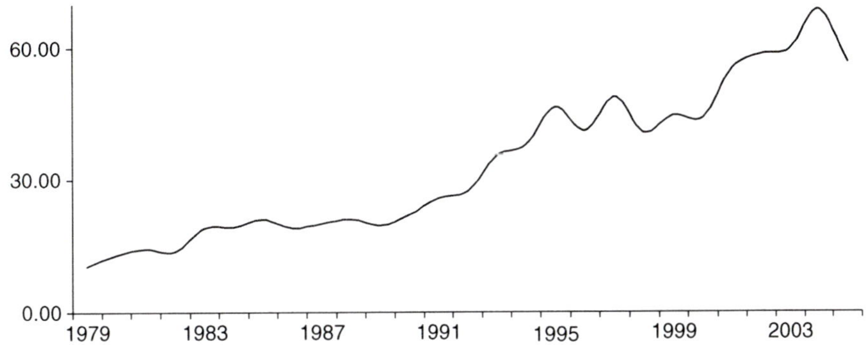

Fig. 5.3 Development of scientific publications on passive houses, 1979–2003
Note: Number of publications on the vertical axis. Figure displays a three-year moving average.

lications in the early 1980s is due to a strong series of papers in *Solar Energy* and *Energy and Buildings* about passive solar buildings. Only a limited number of journals were observed as being active in this field. These basically comprise only *Solar Energy*, *Energy and Buildings* and *Buildings and Environment* which are active in publishing papers in the domain of passive buildings. Over the past 20 years there has been an increase not only in the number of papers, but also the variety of journals in this field.

Figure 5.4 depicts the development of the scientific publications within the SCI using the refined search strategy described in step 4 (2) of Exhibit 5.3. The development shows a comparable pattern: a peak in the early 1980s, a trough in the late 1980s and the early 1990s and a continuous rise since the mid 1990s. Although one can argue that there are no passive house related publications in the early 1980s as the term and the concept were coined in the early or mid 1990s, the development of the passive house concept is clearly based on and makes reference to developments

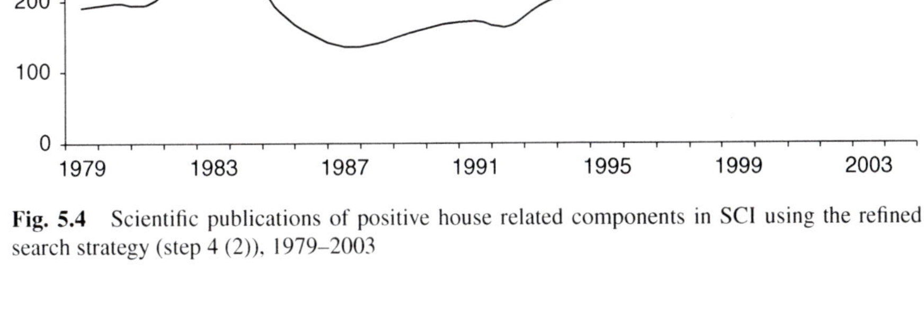

Fig. 5.4 Scientific publications of positive house related components in SCI using the refined search strategy (step 4 (2)), 1979–2003

Exhibit 5.3 Strategy for identifying scientific publications in the context of passive houses

Step 1 Publications which evaluate and briefly describe passive house technologies such as Schnieders and Hermelink (2006), Klingenberg (2005) and Wall (2005) were used to extract a fairly narrow and clear set of English keywords from the text.

Step 2 The extracted keywords were used to identify publications in the SCI, where both the title and abstract were searched. As the narrow set of keywords used for the SCI search and the multitude of components for the passive house made it probable that an essential part of the passive house relevant publications would be overlooked, we decided to broaden the search strategy. The identified publications from the SCI yielded a number of journals where passive house related material is most likely to be published.

We identified the following journals: Energy and Buildings, Solar Energy, Building and Environment, Renewable Energy, Applied Energy, Energy, Applied Thermal Engineering, Energy Policy, Solar Energy Materials and Solar Cells, Heat Recovery Systems and CHP, and Renewable and Sustainable Energy Reviews.

Step 3 This list of journals was subsequently searched using a broader set of keywords to capture all passive house related publications or publications related to energy efficient construction.

Step 4 A natural language processing was performed on the titles and abstracts of the publications identified in step 3. This allowed us to enhance the publication search in two ways:
(1) Selecting only publications with phrases and word combinations which directly indicate passive building technologies enabled us to identify the publications most directly related to passive houses by having the keywords 'passive' and 'house' or 'building' close to each other in the abstract or the title.
(2) Refining the search strategy by the most frequently used technical phrases (top 40) to be applied to the SCI.

in the 1970s and 1980s. The publications identified prior to the mid 1990s clearly indicate the presence of the technological and scientific soil in which the idea and the concept of passive buildings could flourish.

Market Trends

At the end of 2003/beginning of 2004, the Fraunhofer Institute for Solar Energy Systems together with partners carried out market research among architects, manufacturers of prefabricated houses, builders, housing services planners and producers of ventilation appliances to find out how they assessed the development of passive houses and the 3 litre house (Bühring et al., 2004). Whereas only approx. 70 buildings with 120 dwellings of passive house design existed throughout the whole of Germany in 1998, the stock at the end of 2003 was estimated at around 4,000 passive houses, which is equivalent to a growth rate of about 100% per year, see Fig. 5.5. Compared to this, around 6,000 grants for KfW60-houses were approved by the KfW up to the end of 2003. Of the experts surveyed, 43% believed there could be a stock of passive houses of 20,000 and more in 2006, only 16% thought a stock of 30,000 dwellings achievable, as targeted at the time by the

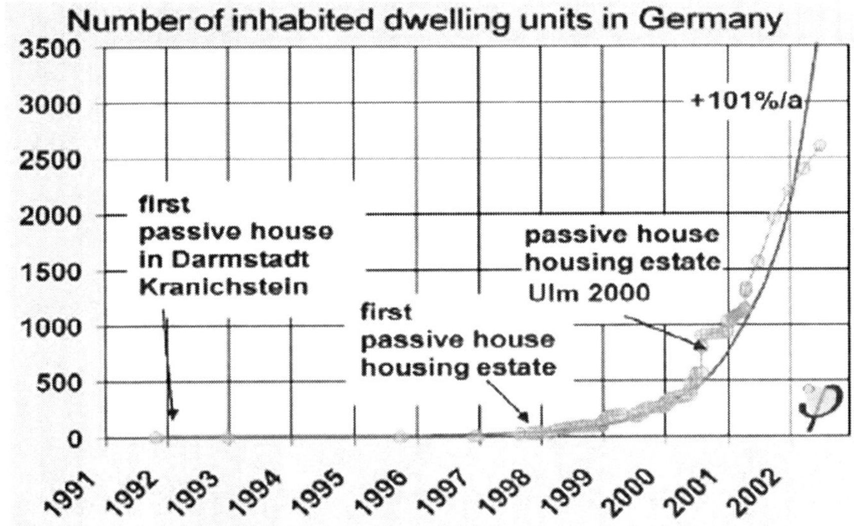

Fig. 5.5 The number of passive houses doubles almost every year, 1992–2003
Source: Schnieders/Hermelink: 2006.

"red-green" coalition agreement. The sector representatives were much more optimistic for the year 2010: here they estimated the share of passive houses among new single family houses (SFH) at almost 20% and almost 13% for multi-family homes (MFH). Architects proved particularly optimistic with regards to the build-up of passive houses, whereas the manufacturers of prefabricated housing estimated only half this penetration on average. Although the experts questioned saw no reason today to build to a poorer energy standard than that of the passive house, for 2010 they estimate that approx. every third house will be built in the 3 litre style and thus that half the new houses built will be of a lower standard.

The Austrian experts questioned in the study are much more optimistic for their country regarding the share of passive houses related to new homes in 2010: they see a share of 28% in SFH and 25% still among MFH. In contrast to this, according to Swiss experts' estimations, these shares in Switzerland will only be 9% and 5% respectively.

Figure 5.5 shows the development over time of the number of passive houses built. This suggests that the number doubles every year in the first 10 years. Although the database only contains a sample of the total number of passive buildings, as the input to the database is voluntary but free of charge, the data depicted shows the overall development of passive house installations in Germany until very recently. After the mid 1990s we observe an increase in passive house installations. From early 2000 onwards, the growth curve becomes quite linear.

Interpreting this development in the context of an s-shaped diffusion curve, one clearly observes that the passive buildings are in the middle of their take-off phase (or rise phase).

To summarize the discussion above: both the indicators of the technological development of passive houses and the indicators of market diffusion suggest that passive houses are in the rise phase or early stage of the diffusion phase of the technology cycle. This assessment is also shared by most of the experts interviewed as part of the innovation system analysis below.

5.3 The Technological Innovation System

The following section sketches the technological innovation system of passive buildings.

5.3.1 The Overall System

It is the overall technological innovation system which is responsible for the design and creation of the passive house and its components. A large part of the system can be associated with the construction industry and the respective innovation system of this sector and its suppliers. The construction sector's innovation system is more comprehensive than the technological innovation system of passive houses and buildings. The technological innovation system comprises a heterogeneous set of actors. Actors from each of the subsystems contribute to the design and generation as well as the diffusion of the passive house. Regulations – as we will see – play a minor role in shaping the innovation system, its actors, their strategies and their relationships.

The following discussion strives to integrate all parts of the innovation system. It does not break down the technological innovation system of the passive house into several innovation subsystems for each of its technical components.

As the technological innovation system as a whole creates the passive house and passive houses are regarded as systems integration innovations,[45] we will elaborate on the effects this type of innovation has on the innovation system and its properties.

The passive house as a systems integration innovation represents a challenge or even a threat to certain established actors in the innovation system. Since the components remain more or less unchanged as far as their core technologies and design principles are concerned, actors focusing on components are not challenged in the first place. The passive house as a systems integration innovation poses a threat to the knowledge base of the actors responsible for integrating components into the final product. Not only their knowledge, but also the procedures, interactions and routines of these actors are threatened by the novelty of the system integration knowledge required by the passive house innovation. For this group of actors, which we have to specify below in more detail, their existing knowledge, procedures and

[45] Systems integration innovation refers to the concept of "architectural innovation" in the sense of Henderson and Clark (1990)

routines may not only be challenged by the new systems integration requirements posed by the passive house; their existing knowledge may actually even be an obstacle to them perceiving and recognizing the new developments in the design of the systems integration. This may create economic advantages for other actors who are faster to grasp the implications of the passive house; it provides opportunities for entrants and incentives for organizational change.

At the same time, the established systems integration knowledge generates doubts among the relevant actors about whether or not passive houses can deliver good and reliable as well as comfortable housing services. The influence of these actors' doubts and their interaction with the demand cannot be overestimated.

To a large degree, the actors in the innovation system of a passive house are also embedded in the technological innovation system of conventional buildings or low-energy buildings. These two represent the main technological alternatives to the passive house.

Alternatives here are a matter of degree rather than of clear cut definition since these alternatives do not deliver identical services to the final customer. Yet they can all be regarded as supplying the basic services expected of a residential building. Even though they are the established product, we also speak of an innovation system when referring to conventional residential buildings. The innovations generated here are most likely to be incremental, developing and reinforcing core concepts in the building components, but leaving the design and the systems integration composition of the building and its construction essentially unchanged.

The actors' activities in the passive house innovation system thus concern switching from the dominant design of residential buildings to the new design of passive houses. This switch has major implications for the actors' knowledge base, strategies and activities. The challenge lies – as alluded to above – in implementing the change in the systems integration design and comprehending the changes required to do so. As the conventional or even the low energy construction of buildings shapes the actors' routines and processes as the dominant design, adopting the new systems integration design requires a reassessment of all routines and the willingness to potentially abandon them. Two factors may hamper this willingness: first, in a complex environment, routines are a way of reducing complexity for actors with bounded rationality. They shorten decision times and provide stability for the actors and the system. Giving up routines potentially increases the instability of the system and the uncertainty for the actors and incurs additional transaction costs. Bounded rational actors – particularly if they are also risk averse – shy away from changing routines. Satisficing behaviour – to proceed as in the past until a certain threshold of mal-performance is passed – is a behavioural pattern which upholds conventional routines and hampers the switch to the new systems integration design.

Second, actors are not isolated from one another in the system; indeed both the design and construction of conventional or passive buildings require the interaction of a large number of heterogeneous actors. Each actor's processes and proceedings are deeply embedded in a network of other actors within the system. The routines, communication and actor's strategies are shaped by the design of the final product: the new systems integration knowledge about how to integrate the components requires new types of interaction and communication. Hence the switch for a single actor from the conventional to the passive house innovation system can only be successful if it is coordinated with other actors in the network. This creates coordination costs which reduce the actors' willingness to change to the new systems integration design. Even if the knowledge about the new systems integration design is freely available and the actor possesses the capabilities to meet the new systems integration requirements, these two factors still represent a barrier for actors in conventional building to switch to the new design.

The passive house cases shows that even though the passive house standard is open and key actors in the field have initiated strong dissemination policies, the acceptance of the passive house remains rather limited when compared to the technological alternatives.

The actors in the conventional residential building sector organize themselves around the product – the system – of a conventional residential building and its technological requirements. The communication channels and the routines of interaction as much as the individual capabilities constitute the systems integration knowledge in the conventional building sector. The passive house as a systems integration innovation challenges this systems integration knowledge.

Take, for example, the interaction of an architect and a firm responsible for the installation of plumbing, heating and ventilation systems. For the conventional construction of a residential building, the architects rely on the expertise and the planning capabilities of the firm. In the design of a residential passive house, the interaction of heating, ventilation, plumbing and possibly sewers is more complex and an additional planning step is required involving specialized engineering firms. The established procedures of collaboration between the architect and the traditional firm have to be modified to adapt to the new design requirements.

The communication channels and the established routines of the conventional building sector create information filters which selectively perceive developments within and outside the system. The information filters are shaped by the technological knowledge and the interaction of the actors in the conventional system and after a certain period of evolution constitute part of the systems integration knowledge of the system.[46] One example of an information filter is the long-time neglect of energy issues – and energy efficiency in particular – in the architects' education curricula. These information filters created blind spots for architects and at the same time opened up a window of opportunity to actors from other disciplines, such as physics, to enter the innovation system and actively shape it by fostering the new systems integration knowledge.

Before actors from the established construction sector can reap the potential benefits from switching to the new systems integration design they have to engage in an intensive learning process to update their knowledge base by the new developments

[46]Of course the system does not have the capacity for mentally representing knowledge of any kind. The term "of the system" is interpreted as "of the actors of the system". We acknowledge that the sum of the actors' knowledge and their interaction create emergent properties – indicating that the system as a whole is more / is capable of more than the mere sum of the actors involved.

and to eliminate the old information filters. This in turn causes old knowledge to become obsolete and incurs costs which have to be borne immediately. The returns, in contrast, are uncertain and potentially in the distant future.

The uncertainty is mainly generated by three factors. (1) The consumers' demand for dwelling units built in accordance with the passive house standard today – as will be discussed below – is strongly influenced by the expected development of energy prices. The actors' bounded rational deliberation of the potential scenarios seems to induce a rather positive estimate of the future energy price developments. (2) The consumer's expectation of his or her future disposable income also frames the possibility of being able to afford dwelling units based on passive house technologies. (3) In addition to this, the development of consumer preferences for certain characteristics cannot be predicted. They are assumed to be rather persistent over time.

The established routines, the obstacles to switching, the required learning and the uncertainties in the composition of the final demand generate three different strategies among actors.

The first strategy is complete commitment to the innovation system of conventional residential buildings. Innovation and improvement strategies focus on incremental innovations.

The second strategies is complete commitment to the new systems integration model of passive houses. The innovation strategy of the actors is targeted towards managing the systems integration change and changing the actors' internal procedures and processes so as to fulfil the requirements for successfully contributing to the generation and diffusion of passive house technologies.

The third strategy observed is a blend of the first and second strategy. Actors try to be present in both the conventional building innovation system as well as in the passive house innovation system. This strategy can have two underlying causes. One is that the actor tries to diversify his or her portfolio of activities based on Tobin's notion of not "putting all the eggs in one basket". The other is that the actor is currently in transition from strategy one to strategy two and in the process of switching his/her focus from conventional to passive buildings.

In the discussion below we will try to elaborate on the actors in the passive house innovation system in more detail. To do so we will group the actors according to their respective subsystems.

5.3.2 *The Various Elements of the Innovation System and Their Roles*

The passive house innovation system has undergone a fairly dynamic development over the past decade. Public interest, as exemplified in the passive house symposia (the first of which took place in late 1996), shows an initial rise and then, since the early 2000s, stabilization. Commercial activities, as exemplified by the participation of exhibitors in the passive house congresses, are quite cyclical in nature. The number of exhibitors drops if the exhibitions are outside Germany as, e.g. in 2002 and 2004 where the congress was held in Basel, Switzerland, and Krems, Austria,

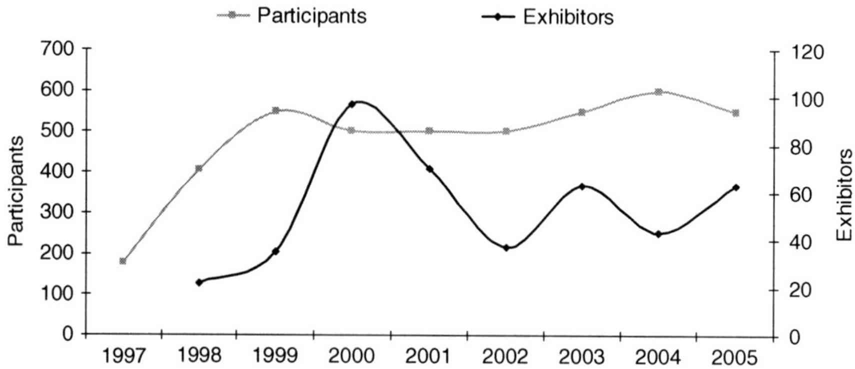

Fig. 5.6 Participation in passive house exhibitions and symposia, 1997–2005

respectively. Figure 5.6 illustrates the development of both participants (around 500 per event since 1999) and exhibitors (40–60) per event.

5.3.2.1 Industry

The industrial subsystem of the technological innovation system for passive house technologies relates to the development and production of components such as ventilation systems, windows, insulation etc. It also encompasses companies delivering passive houses to the final customer such as the manufacturers of prefabricated houses and buildings.

Strictly speaking, each of the producers of passive house components belongs to a separate technological innovation subsystem. These cannot be analyzed and discussed here. Instead we attempt to provide an overview of the central actors in the system by selectively focussing on certain aspects of the overall innovation system of passive houses and their components.

Figure 5.7 shows the development in the number of companies mentioning "passive houses" in the description of their field of business when officially registering the company. With 0–10, the case figures are too low for a statistically sound statement, but they tentatively support the hypothesis. The trend over time is more or less synchronous with market penetration, if the sharp drop in registration numbers after the year 2003 is disregarded. This could either be an artefact due to the time lag in registrations or – and this seems more probable – to the fact that established companies have increasingly extended their business activities in the direction of passive houses. The strong recent growth in passive houses could not otherwise have been possible.

The regional distribution of the above mentioned company registrations (Fig. 5.8) appears to refute the starting hypothesis that companies have mainly settled in the area around Darmstadt to be in the vicinity of the Passivhaus Institute. Indeed, the much smaller number of entries in region 5 seems to indicate that the market is already well covered in this area, but the few observations do not allow for a robust statistical backing of this finding.

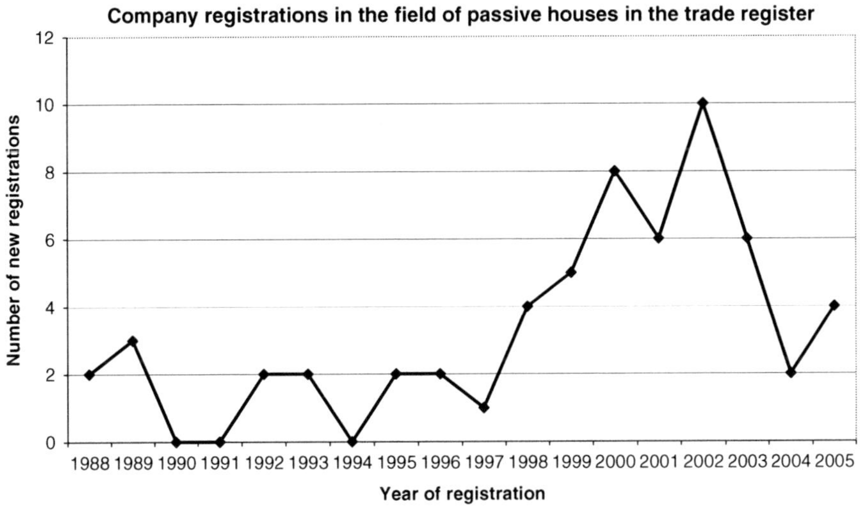

Fig. 5.7 Year of company registration for companies active in the field of passive houses and buildings, 1988–2005

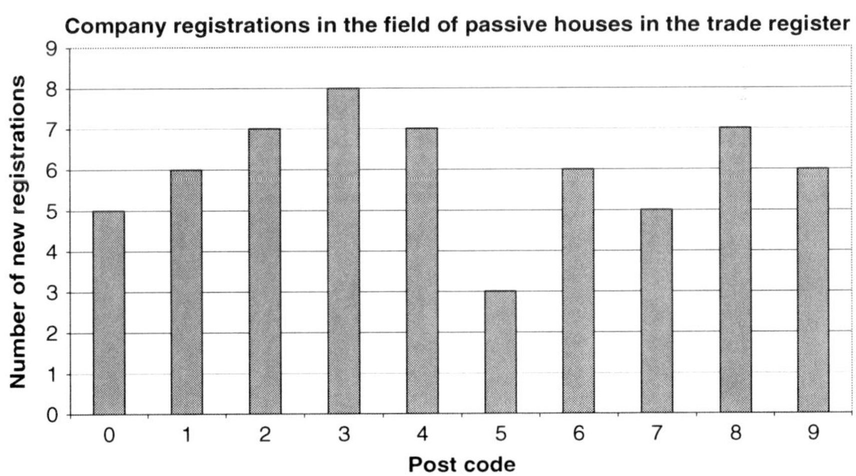

Fig. 5.8 Geographical distribution of company registrations in the passive house sector between 1998 and 2005

Exhibit 5.4 Passive house projects database

In addition to the interviews and the literature study which delivered an essential part of the information compiled in this section, the internet-based database for passive house projects supplied valuable information which could otherwise not have been obtained in this degree of detail (www.passivhausprojekte.de).

The database is operated by Passivhaus Dienstleistungs GmbH, a subsidiary of the Passivhaus Institute, and allows owners of passive buildings to publish information on their building in a standardized internet database.

The information used here from this database only relates to those residential passive buildings built in Germany which are recorded in the database dating from 22 October 2005.

Entries to the database are on a voluntary basis. The whole database contains some 700 entries, 602 of which refer to residential buildings in Germany.

Components of Passive Houses

Let us exemplify the industrial structure by using the passive house projects database (see Exhibit 5.4) and by looking at the ventilation and air conditioning systems actually used in passive residential buildings (see Fig. 5.9). Two dominant suppliers of ventilation and air conditioning components can be identified, each with a market share of more than 30% (see Exhibit 5.5). None of the remaining producers has a market share of more than 4%. Thus, there is a highly concentrated industry delivering ventilation components compatible with passive house requirements. The major actors are quite young or have recently restructured their operations to better serve the markets.

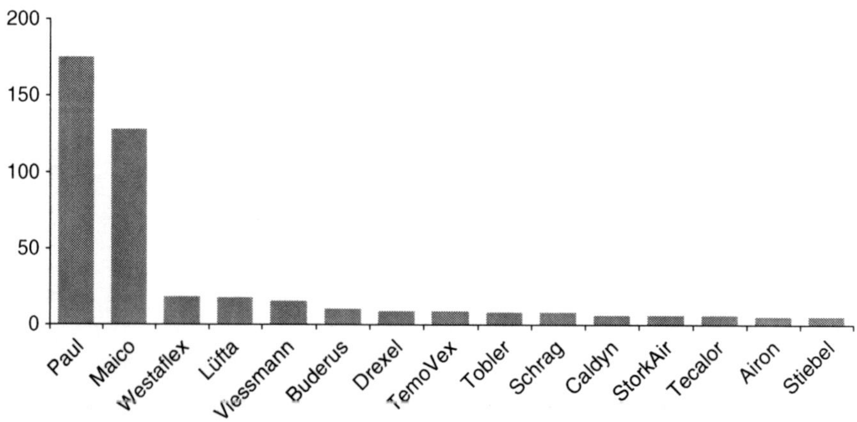

Fig. 5.9 Producers of ventilation (air conditioning) systems used
Note: The diagram is based on information obtained from the description of realized passive house projects in www.passivhausprojekte.de. For 445 of the 602 passive house projects, it was possible to identify the make of the air conditioning and ventilation system. The distribution shown here excludes all producers with a share of less than 1%.

Exhibit 5.5 Leading suppliers of ventilation and air conditioning components

The leading company supplying ventilation and air conditioning systems and components is Paul Wärmerückgewinnung GmbH. It has been in operation since late 1994. It is a small SME with leading innovations in its field and also a wide range of highly-efficient ventilation systems for residential buildings.

Since 2002, MAICO, a heating, ventilation, air conditioning and refrigeration systems specialist based in Villingen-Schwenningen has joined forces in a strategic alliance with Fresh ventilation systems from Eisdorf. The main area of activity is the consulting, supply and training of all actors involved in the construction of high energy efficiency buildings. The supported actors range from tradesmen to architects and even extend to the interested end-user of the heating and ventilation system. The alliance is strategically oriented towards the type of building ventilation which is anticipated to be standard in the near future in Germany. The activity within the passive house sector also seems to act as a vehicle to unleash learning effects utilised later in the area of anticipated future developments.

The strategic interaction of the leading suppliers of components with architects and the contractors and fitters responsible for planning and implementing the ventilation systems increases the diffusion of the systems integration knowledge about how the components are integrated in the whole passive house system.

Prefabricated Houses

As mentioned above, the systems integration nature of the passive house innovation represents a major challenge to those actors responsible for the integration of the individual components, planning and design. The manufacturers of prefabricated buildings are in a contradictory position. First, as integrators of components it falls in their domain to ensure the produced building meets the passive house standard. Through prefabrication, these producers also considerably reduce the work done at the construction site. This makes it possible for them to establish procedures and routines to meet the passive house requirements on a quasi-industrial basis at their production sites. However, it also shows that precisely these existing routines and production methods create problems when assembling the prefabricated pieces of the building in such a way as to meet the air tightness standards. Furthermore, even though passive house production may be an option for some manufacturers of prefabricated buildings, the long and costly systems integration construction and design process – requiring new and modified ways of calculation and planning – incurs relatively high costs before the first unit can ever be built and sold.

This is exacerbated by the fact that not only do financial constraints have to be overcome, but mental reservations also have to be taken into account. In a gross simplification we could state that meeting energy efficiency targets and potential monetary profits are the main driving forces behind house manufacturers embarking on the passive house route. Yet, energy efficiency plays or has played a minor role

– if any at all – in the education of most architects and construction engineers; hence, the energy efficiency of a building is not on most planners' priority list when planning a building. The rather low level of basic education in energy efficiency which generations of planners have experienced during their training makes it even harder and more expensive in terms of time costs to develop the demanding planning capabilities required to meet the passive house standard.

Profits – as the second major driving force – are highly uncertain as the demand for passive residential buildings can hardly be predicted since this depends on economic, societal and political factors. Amortisation of the cost and benefiting from the experience cannot be expected for the majority of manufacturers as most of them have (up to now) realized less than 10 units. Only a few have experience with more than 20 units.

Where the manufacturers of prefabricated buildings conduct R&D they do so mostly in collaboration with suppliers, customers and architects. In addition, although their main challenge is the systems integration dimension of the innovation, some manufacturers develop new and improved components such as insulated wall systems etc. R&D hardly involves research organizations since – given the potential profit margin – the costs of formally collaborating with institutes from the Fraunhofer association are too high for the manufacturers.

Although they may be hampered by the facts discussed above, some manufacturers of prefabricated buildings still regard energy efficiency as a competitive factor for their products. However, this esteem for energy efficiency has not progressed far enough for the manufacturers to completely commit themselves to passive house buildings. Nevertheless, if we extrapolate the corporate strategies and the rumours in the field, this may well be within reach.

Architects

Although we can observe a dynamic interest in passive buildings and positive prospects among the actors involved, the passive house – in relation to all realized buildings – still remains a niche product. Architects with efficiency and experience in designing and planning passive buildings also suffer from the low number of realized units per architect. The required development in systems integration knowledge remains almost constant for any given number of units. Detailed knowledge on the planning and design of foundations, thermal bridges and their avoidance, the sizing of the ventilation system etc. is required for the first passive house built by any planner. The necessary novel workflow and the required precision of work not only challenge architects' established procedures and routines; they also challenge their communication principles and their cooperation routines.

As a passive building requires the seamless integration of several (complex) components and a reliable and knowledgeable site manager, the working processes on the construction site have to be adapted to meet the passive house standard (see e.g. Horn, 2005). The quality of execution is a crucial factor in building a passive house. New ways of quality assurance and quality management demand new procedures such as the blower-door test. Beyond updating the routines and the processes for planning and erecting a passive building, this requires the architect to communicate new tasks which have to be paid for by the customer. There are strong incentives for the architect to avoid these discussions with hesitant clients by cutting short the requirements for a passive building and suggesting a low-energy building instead.

Although architects were doubtful about the strict technical requirements (e.g. relatively voluminous insulation etc.) for passive buildings early on, many have since realized that passive buildings can also be designed in an aesthetic way (Plöderl, 2005). Documentation material by the IG Passivhaus (www.igpassivhaus.de; www.igpassivhaus.at; www.igpassivhaus.ch) vividly demonstrates that energy efficiency and aesthetic buildings are not contradictory principles. Indeed passive buildings may even have design advantages since they do not require radiators in the buildings; ceiling to floor glazing can be implemented without the obstruction of radiators in front of the windows.

5.3.2.2 Construction Site Companies and Craftsmen

Since the manufacturers of prefabricated houses are keeping out of the business with passive houses to a large extent, most of these homes were built in the classical way predominantly by hand on site, although a trend towards prefabricated elements is becoming apparent. Whereas the supervision by the site manager may be relatively lax for conventional houses since errors during construction are hard to spot later on, the passive house standard can only be achieved through exact, error-free construction. Since this represents a huge challenge for the builders especially with relation to air-tightness, this is tested using the blower-door test before each final acceptance of the construction work. This strict quality control must trigger a certain aversion to constructing passive houses among many building companies and trades people. However, the general slump in the construction industry which has being going on for some time could have encouraged some companies to become active in the passive house sector due to a lack of conventional building contracts or due to a high quality standard which also qualifies them to build passive houses and thus distinguish themselves from the competition.

5.3.2.3 Public and Non-Profit Organizations

Public research plays a role in the construction and design of passive buildings. However the major role of public research organizations is related to the research and development of individual components for the passive buildings such as heat exchange pumps, wall and window systems etc. The public research organizations seem less relevant when it comes to the improvement or the dissemination of the systems integration knowledge dimension.

Given the current lack of passive house issues in university and college (FH) education (Plöderl, 2005), strong leverage could be achieved by incorporating passive house relevant topics in the education and training of architects and civil engineers on a standardized and compulsory basis.

Publicly funded research also contributes to the development and dissemination of the passive house standard. For example, the EU funded project "Cost efficient passive houses as European standard" scientifically evaluates the design of and the comfort achieved by passive houses all over Europe. This project was co-funded by the Hessian Ministry of Environment, Agriculture and Forests.

Interest Organisations and Other Corporate Actors

There are very few large players in the passive house innovation system. A large share of the manufacturers of components as well as of the building as a whole is made up of small and medium sized enterprises, many of which have less than 10 employees. The larger actors are not completely committed to the passive building innovation system. Both the small size of the committed companies involved and the reduced level of commitment in the part of the larger companies affect the possible formation of lobby organizations to push the passive building at the appropriate political level (Plöderl, 2005).

5.3.2.4 Passivhaus Institute

Wolfgang Feist, his Passivhaus Institute and its subsidiaries are the central actors in the German passive house innovation system. The Passivhaus Institute was founded in 1996 by Dr. Wolfgang Feist as an independent research institute. It accumulates multidisciplinary expertise around the passive house and employs physicists, mathematicians, civil, mechanical and environmental engineers.

The Passivhaus Institute offers advice and support on the planning of passive houses as well as the development and optimisation of components suitable for passive houses. The staff conduct scientific surveys and quality certification for passive houses and their components. They compute interior air flows, conduct daylight simulations and offer tracer gas measurements, thermography and tests of building air tightness.

The Passivhaus Institute defines a private, yet open, passive house standard and offers certification services for suppliers where certain components are certified to comply with the passive house standard. The PHI also offers the certification of completed passive houses which meet the passive house standard. The certification plays an essential role in mitigating the uncertainties described above. It reduces uncertainty about the performance of passive house components and reduces the uncertainty of customers as to whether the building they pay for really meets the passive house requirements.

The Passivhaus Institute maintains an extensive network with all types of partners in the innovation system. It conducts an open policy of collaboration. It does not hold IPR for "passive house" or related technologies, only a trademark for the visual sign of the passive house certificate. For further details refer to http://www.passiv.de

5.3.2.5 Research Institutes

Fraunhofer Institute for Building Physics, IBP

This Institute is active in the design, consultation, practical measuring and determination of low-energy, low entropy, low exergy and zero heating energy houses and housing schemes and the calculation and realisation of energetic redevelopment concepts for existing buildings. Projects include the development and testing of facades, heating, ventilation, solar, hybrid, storage and energy-supply systems for practical use in buildings, as well as drawing up and developing total energy balances (including ecological balance share) of buildings and heat supply systems. Analyses and assessments of energy potentials in new and existing buildings are also conducted as are calculations and measurements of air flows in rooms and large halls (atria). The temperature behaviour of buildings in summer is assessed and calculations and measurements made of the lighting and daylight provision in buildings. The development and updating of computer-aided planning instruments and information systems is another area of activity. The Institute is further involved in the coordination of integral international research and standardization projects (www.ibp.fhg.de).

Fraunhofer Institute for Solar Energy Systems, ISE

This institute is active in the following areas (www.ise.fhg.de):

- Building concepts, simulation and control: simulations performed during the early planning phase optimally synergize the building envelope, technical building equipment and climatic factors at the site thus detecting potential cost reductions, areas for improvement or optimization possibilities at an early stage. Particular emphasis is placed on lighting simulations, indoor comfort analyses and the detailed modelling of complex energy supply systems.
- Facades and windows: optimisation and testing of components and facades; evaluation of the energy efficiency and development of adapted control strategies for dynamically changeable facades. Coatings and microstructures are used to equip surfaces with special properties, e.g. infrared reflection, anti-reflective and self-cleaning effects. Switchable glazing adapts the visual contact and solar control according to the users' wishes.
- Heating, ventilation and air conditioning technology: the development and measurement of components, systems and controls; use of customised simulation programmes to develop comprehensive software solutions.
- Monitoring and demonstration projects: the design, instalment and operation of complete measurement systems of all dimensions. Results are compared with calculated values (SIMULATION) or measured results from comparable buildings, plants and photovoltaic systems. Monitoring is an essential component of demonstration projects and supports quality control of components and systems, e.g. meeting planning specifications in buildings or supplying guaranteed yields from thermal or photovoltaic systems. Recommendations for the rational use of

energy and the application of renewable energy sources can be derived from the results. At the same time, the results of demonstration projects are used to exploit further potential for optimising systems technology and improving building quality.

5.3.2.6 Institut Wohnen und Umwelt GmbH (Institute for Housing and Environment)

The Institute – legally structured as a limited liability company – is a research institution of the Federal state of Hesse and the city of Darmstadt. The task of the Institute is to research, in interdisciplinary cooperation, present forms of housing and living conditions. A particular goal is to contribute to improving the housing conditions of socially weaker groups. Moreover, the possibilities of minimized, efficient, environmentally and socially acceptable energy use are to be investigated. The Institute has played a pioneering role in introducing the low-energy house standard in Germany (www.iwu.de).

5.3.2.7 Research Programme "Energieoptimiertes Bauen[47]" (ENOB)

In the "Energieoptimiertes Bauen" (ENOB) programme of the Federal Ministry of Economics and Technology (BMWi), the promotion of R&D projects on efficient ventilation systems and concepts, i.e. components for passive houses, plays an important role alongside the pilot and demonstration projects which were and are being promoted and many of which comply with the passive house standard. Other main fields of support within ENOB include highly efficient thermal insulation and efficient windows and solar protection systems – all of which are directly relevant for passive houses.

Experiences from the programme "Energetische Sanierung im Altbaubestand"[48] (EnSan) have an indirect effect since many components from the passive house field are also used here and experiences from the scientifically monitored demonstration projects are fed back to the component producers and developers.

5.3.2.8 Cost Efficient Passive Houses as European Standards (CEPHEUS)

CEPHEUS is a project within the THERMIE programme of the European Commission carried out by the DG Transport and Energy. The project ran from January 1998 to December 2001.

---

[47] "Energy optimised building"

[48] "Energy-related refurbishment in older buildings"

Exhibit 5.6 Goals of the CEPHEUS project

The construction and scientific evaluation of approx. 250 housing units built to Passive House standards in five European countries with the following goals:

- To demonstrate the technical feasibility (in terms of achieving the targeted energy performance indexes) at low extra cost (target: compensation of extra investment cost by cost savings during use) for an array of different buildings and designs implemented by architects and developers in a variety of European countries;
- To study investor-purchaser acceptance and user behaviour under real-world conditions for a representative range of buildings;
- To test the implementability of the Passive House quality standard throughout Europe with regard to cost-efficient planning and construction;
- To provide opportunities for the general and the informed public to experience the Passive House standard hands-on at several sites in Europe;
- To give development impulses for the design of energy- and cost-efficient buildings and for the further development and accelerated market introduction of individual, innovative technologies compliant with Passive House standards;
- To create the preconditions for the broad market introduction of cost-efficient Passive Houses;
- To illustrate, based on the concrete example of the Hannover-Kronsberg subproject, the potential of the Passive House standard to provide a basis on which it is possible to meet the energy requirements of new housing in a manner that is both cost-efficient and, taken over the whole year, produces zero greenhouse gas emissions (climate neutrality criterion);
- To present this sustainable – fully primary energy and climate neutral – approach to the energy supply of new housing developments at the EXPO 2000 World Exposition in Hanover, in conjunction with all CEPHEUS subprojects.

CEPHEUS, an European demonstration project, resulted in the construction of about 250 housing units which meet the passive house standard in five European countries (see Fig. 5.10), with parallel scientific support and the evaluation of building operation through systematic measurement programmes (see Exhibit).

5.3.2.9 Intermediaries, Financial and Further Institutions

One of the most visible financial institutions involved in funding passive houses is the German Kreditanstalt für Wiederaufbau – KfW banking group (German Bank for Reconstruction). Founded in the wake of WWII, its main target has been to

Fig. 5.10 Location of CEPHEUS buildings
Source: Schnieders and Hermelink, 2006.

support the reconstruction of the German economy. The initial equity capital was supplied by the Marshal Plan. The federal government currently holds 80% of the KfW shares and the German Bundesländer jointly hold 20%. Ever since it was founded, the KfW's objectives have adapted to national and international challenges and now cover financial support to developing countries and financial support to the transition of former socialist economies.

Among others, the KfW, the Kreditanstalt für Wiederaufbau, gives loans for housing construction and modernization and energy conservation. The programmes specifically for passive houses (KfW40 houses) and 3 litre houses (KfW60 houses) are described in Sect. 5.3.3 below.

5.3.2.10 Demand

In the initial phase of the passive house the demand came predominantly from consumers driven by ecological motives who had an above average willingness and ability to pay.

The current demand for passive residential buildings comes from households with an above average income and a good educational background. The average age of passive house customers increased over the past decade (Horn, 2005). Rasmussen and Grocholl (2005) see the cultural creative milieu as behind the major demand for passive residential buildings. Despite having good prior information on consumption and investment goods, emotional consumption decisions and rational decisions rank equally in their estimation. The cultural creative milieu is more likely to be affected by ecological issues and sustainability questions.

Based on previous experience, all estimates indicate that – everything else being equal – passive house buildings built in 2006 are about 5% to 15% more expensive than conventional buildings.[49] The possibility for trade-offs, however – in new and expensive kitchen furnishings, bathroom equipment or floor coverings – harbours the potential to reduce the construction and furnishing costs of a passive house to the level of a conventional building.

The **higher price** is the main reason why potential customers may shy away from passive houses. Other reasons for not considering the passive house standard when constructing new residential buildings are more deeply rooted in customers' preferences. Potential customers simply cannot imagine living in a *building without a heating system.* People seem to cherish the possibility of *"lighting a fire to keep the cave warm"*. Other preconceptions, e.g. that *windows cannot be opened in a passive building,* also contribute to lowering the demand for passive houses. Mainly, however, it is the higher investment cost involved. Even though amortization times are predicted to be within 25 years, the *additional costs still have to be borne* when financing a passive house building, even if its overall life cycle costs are considerably lower than for a conventional building.

From a pricing point of view, the higher cost of the passive building is matched by its superior performance. As mentioned above, the higher cost is not always matched by a higher consumers willingness to pay, as the willingness to pay always subsumes the ability to pay. However, for builders, the ability to pay for a passive house does not always create the corresponding willingness or desire to pay as the house does not distinguish itself visibly from conventional buildings. The passive building does *not* seem to have the *potential to become a life-style product.* Hence, it seeks buyers in the narrow niches of the housing market where ecological consciousness and a sufficiently high disposable income coincide.

The large number of building variants which are more energy-efficient than the current building stock or current conventional buildings contribute to the confusion of potential customers.

---

[49] See e.g. Haum and Nill (2004) p. 9.

Essentially the unique selling point of a passive house in comparison to a conventional building or a low energy house is its greater independence of future oil and energy prices. The argument thus draws strongly on the builder's/owner's risk aversion with regard to energy prices. However, the same builder/owner, who is risk averse with regard to energy prices, also has to be pro-risk to consider living without a heating system, and to accept the automated ventilation system, which is not very common in conventional German residential buildings. It is hard to find both these traits in combination.

5.3.3 Public Policy

5.3.3.1 Funding by the German Länder

The German Bundesländer usually grant loans for social reasons for general housing renovation. Programmes to financially support the energy-related modernization of older buildings are only known in Baden-Württemberg, North Rhine-Westphalia and Rhineland-Palatinate. The Bundesländer refer to the KfW programmes, see below.

5.3.3.2 KfW – German Bank for Reconstruction

Having introduced the German bank for reconstruction (KfW), a brief review of its financing activities in the context of passive buildings is given below: at the beginning of 2005, the KfW reorganized the subsidy programs for new buildings and reconstruction to make them more transparent for the potential beneficiaries.

Renovation and Restoration of Buildings

Four funding mechanism options are available for reconstruction depending on the degree of energy saving investments as a share of total renovation investments in the building. For energy saving investments which amount to less than 1/3 of the total renovation expenditure, the *standard* option is available. The *mix* option is available for a share of energy savings investments which is larger than 1/3 of the total investments. The *Öko-Plus* (*eco-plus*) option is specially designed for the case of making a building more energy efficient measured by the adjusted primary energy consumption. It provides incentives for insulation of the building envelope and investment in heating technologies which are based on renewable energy sources. Even a heating system based on fossil fuels may be subsidized as long as it is combined with solar panels. In addition to this, heat pumps, ventilation systems and biomass combustion systems can be subsidized if the investment in energy saving technologies amounts to 2/3 or more of the total investment. The most attractive option seems to be the CO_2 building reconstruction programme. The subsidy targets the reduction of CO_2 emissions and ties the size of the subsidy to the achieved level of reduction. A reduction of at least $30\,kg/m^2$ is required for participation in the subsidy programme.

At a reduction of 40 kg/m^2 or more, the maximum loan of 250 Euro/m^2 can be granted. An additional restriction is that the renovated building originates from 1977 or earlier. Partial abatement of the debt makes the schemes even more attractive.

New Buildings

The option *Ökologisch Bauen* (*build ecologically*) funds energy-efficient buildings regardless of whether they are inhabited by the owner or leased out. The size of the subsidy hinges on the primary energy demand of the building and varies between 30,000 and 50,000 Euro.

If the primary energy demand of the building is less than 60 kWh/(m^2a) and the building's transmission heat loss does not exceed 70% of the level stipulated in the EnEV then a loan of up to 30,000 Euro can be granted with reduced interest rates.

If the primary energy demand of the building is less than 40 kWh/(m^2a) and the building's transmission heat loss does not exceed 55% of the level given in the EnEV then a loan of up to 50,000 Euro can be granted with reduced interest rates. Passive houses usually fall into this category. Hence they are eligible for a loan of up to 50,000 Euro. The funding received through the *Ökologisch Bauen* scheme can be combined with any other funding scheme available.

A second strong funding scheme, which may apply to some passive houses, is the solar electricity funding scheme. If the building is equipped with photo-voltaic panels it may be eligible for a loan of up to 50,000 Euro at a reduced interest rate.

In addition to the direct support for investments, the federal government also subsidizes energy consulting services to assist builders in their decisions about energy efficiency and related investment and planning issues.

The application procedure has turned out to be a clear obstacle to the KfW support. Funds must be applied for by interested parties via their own banks. Since these banks only receive a relatively low fee for processing requests and are not able to provide the loan themselves, they really have no interest in handling this transaction. As a result the banks do not refer to the programmes and indeed several cases are known in which the bank was not prepared to process a KfW loan.

5.3.4 Regulation

As discussed above, the passive house standard is a purely private standard with a privately organized certification system. It is not maintained by the German standardisation organisation (DIN) or the International Standardisation Organisation (ISO). Nevertheless, the passive house standard is not set individually but represents a general rule governed by the laws of physics and hygiene.

The introduction of the EnEV seemed to have an influence on the development and especially on the dissemination of passive houses, although the direction of the influence is not entirely clear. On the one hand, the EnEV has stirred the awareness of builders about energy consumption having a direct influence as one of the unique selling points for the passive house: its energy efficiency and independence of future energy prices. On the other hand, the computational scheme proposed by the EnEV to estimate the energy consumption of a building is somewhat counterproductive to the targets of passive houses. The EnEV implicitly allows for a trade-off between energy saving investments such as wall insulation and the use of renewable energy sources for heating and preparation of hot water. This trade-off runs counter to the insight that, without proper energy saving measures, the overall energy consumption of buildings can hardly be covered by renewable energy sources such as wood pellets.

The different ways of computing energy consumption and labelling energy consumption do not contribute to transparency for potential builders.

5.3.5 Collaboration

Rigorous planning of the integration of the components and the design of the building requires intense collaboration for all partners in the construction process. Collaboration is required for quality assurance and the seamless integration of all phases of construction.

5.4 Implications for Public Funding and Intervention

5.4.1 Bottlenecks and Opportunities

Having located the passive house in the upward phase of the technology cycle somewhere between the rise and the diffusion phase, we focus our discussion of the bottlenecks and the opportunities on the underlying economic causes. We identify bottlenecks both in supply and demand. Supply is strongly challenged by the systems integration novelty implied by the passive house design. Demand is hampered by the comparably high pricing of passive buildings, their incompatibility with certain consumer preferences and behaviour, and the proximity of close variants of low energy buildings.

The realization of passive house projects has reached an impressive number and major cost reductions were achieved in the early phase of the development. However, additional cost savings through – for example – scale effects have been rather modest (Haum & Nill, 2004). A potential reason for this is the quasi monopolistic market structure which emerges early in a technology cycle and may persist up until the diffusion phase. Reduced competition results in higher prices for certain components. Despite not having conducted an extensive market structure analysis for all the passive house components, there are indications that some components such as windows and window frames in particular may generate substantial profits for the producers. Also the revealed willingness to pay by passive house customers reduces the incentives for manufacturers to engage in price competition. This is exacerbated by the fact that passive house customers tend to want up-market products such as wooden windows which also reduces the competitive pressure for those niche suppliers.

The polypolistic structure of the system integrators where only few have extensive experience with the generation of passive buildings also reduces the chance to realize scale effects.

Factors persistently hampering the widespread implementation of the passive building standard are (1) the lack of information among potential builders, (2) lack of experience among system integrators and architects in particular, (3) comparably low energy prices (although this issue has changed a bit during the past 18 months), (4) regulation issues, (5) lack of experience of trades people, (6) costs, (7) other competing technologies, e.g. district heating (the objective of district heating companies is an energy density in their area which is not too low – and these companies are quite strong). Hampering factors which in the long run will be of decreasing importance are: (1) quality issues and technical problems, (2) preconceptions of potential builders, and (3) construction costs (see Haum & Nill, 2004).

As mentioned above, the scale based quality improvements and cost reduction learning effects among system integrators do not exert a strong influence in a rather decentralized market with a polypolistic supply structure.

The final demand for passive houses is hampered by the rather peculiar combination of facts. First, apart from ecological considerations a passive house reduces its inhabitants' economic dependence on future energy price increases relative to conventional or low energy buildings. To achieve this independence, the passive house demand has to pay a premium on the construction price relative to conventional or low energy buildings. To pay a premium today to avoid risks in the future is particularly appealing to risk averse actors. Hence residential buildings based on the passive house standard are targeted towards risk-averse builders. On the other hand, building a passive house requires a considerable amount of risk taking as the lack of experience among architects, planners, and workmen has been identified as a major hampering factor. Also taking the risk of running into technical problems and ending up with a suboptimal quality of the final building requires the builders to take risks. In addition passive buildings require the flexibility and openness of builders to change some of their behavioural patterns and their attitudes towards certain characteristics of the dwelling. This also implies that passive houses require inhabitants who are at least risk neutral. Summarizing the discussion, we see that the passive house customer has to be risk averse from a financial perspective but risk neutral at the same time, which may reduce the number of potential clients.

On the other hand, there is an increasing desire of many consumers to be "autarkic" with regard to energy in view of rising energy prices. This self-sufficiency is achievable chiefly by combining passive houses with renewable energy sources. At the same time the number of passive houses is increasing so that they are losing their aura of being something exotic and predominantly "success stories" are getting through to a wider audience due to the constant progress made in technology and experience gained in system integration. There is therefore likely to be an expansion of potential customers beyond that common today with the consequence that not only will more cost-conscious buyers enter the market, but that it should also be possible at last to take the step from producing small series to mass production of components with the corresponding cost reduction effects.

Regional or local building codes do not fully support the construction of passive buildings as these may forbid the strict south-orientation of a building. It is shown that fully functional passive buildings can be built without strong orientation of the building towards the south, even shaded locations are possible. However, deviating from a strict south orientation requires considerably greater experience of the architects and planners and implementers involved. In addition it may increase the building costs. An extensive report on a passive house facing north can be found in Feist, Peper, and Pfluger (2004).

5.4.2 Policy Recommendations

Financing and cost issues constitute the crucial factor in the decision for or against a passive house in most cases. Some basic approaches exist to target the demand, which have to be employed simultaneously to be successful.

The willingness and – above all – the ability to pay for the higher initial building costs have to be increased. Appropriate financing subsidies and interest reduced loans including no payback in the first years etc. should be in place to bridge the financial gap. Appropriate accounting for the reduced life-cycle costs of the building has to be ensured. Increasing the customers' ability to pay for housing, however, does not drive down the unit cost immediately. It may not even set the incentives to do so.

Public procurement of passive buildings can have two advantageous effects. First, it may increase the number of units demanded and hence induce scale effects with the supply of passive house components. Second, through a strict bidding process it may reduce the profits of passive house suppliers and suppliers of its components theoretically to an almost competitive level. To the best of our knowledge, public procurement would be an effective tool to foster the diffusion of passive houses as the procuring authority can fully specify the technical characteristics of the building. If these are fulfilled, the authority then makes a decision based on price. The accusation against innovation policy measures utilising the demand side cannot be brought forth against the procedure suggested here. Public procurement of passive buildings builds on an established set of technical characteristics, which – as vividly illustrated in the passive house database (http://www.passivhausprojekte.de/) – can be achieved by a number of different configurations using a large variety of components from different suppliers. Hence, procurement of passive buildings does not favour certain suppliers. From a competitive policy point of view this seems to be unproblematic.

The risk and the uncertainty about the amortization of the passive building and its technical characteristics call for novel approaches. In combination with the suggested public procurement, one could think about new private public partnerships to launch new models of operation and financing for passive buildings. The underlying idea would be to separate ownership, financing and use of the building to enable even risk averse users to live in a highly energy efficient building such as a passive house. The initial investments for a passive house could be repaid using the energy cost savings. One could also think about a financial instrument which takes account of the fact that investing in passive house technologies is the same as betting on increasing energy prices for private households. The risk involved in this bet could be reduced by appropriate financial instruments.

Our argument for financial support for passive house demand is also backed by the findings of Bühring et al. (2004) documented in Table 5.1. For passive house suppliers, suppliers of passive house components and system integrators, financial subsidies are essential to foster the development and the demand for passive houses. The demand side captured by the property developers also argues in a financial context, where saved heating costs are as important as financial subsidies (see bold figures in Table 5.1).

Arguing in favour of financial support for the demand side in no way rules out the option of funding the development of new technologies, but rather the opposite. The arguments given above highlight characteristics of the technologies which can contribute to the widespread diffusion of passive buildings. Technologies which will contribute to the diffusion of the passive house have to target the hampering factors directly.

Only technologies which have the long-term prospect of reducing the construction costs of passive buildings support the diffusion of the passive house. However, when deciding on supporting a certain technology for the passive house it has to be borne in mind that it is not the cost of the individual component which matters, it is the whole integrated system of a passive building. This reduces the effect that cost reduction in a specific component will have on the overall construction cost. It should also be noted that alternative technologies such as low energy houses based on the EnEV allow for some trade-off or substitution of certain components and technologies. The passive house standard does not incorporate such a comparable trade-off.

On the other hand, due to its high standards, the passive house approach is driving the further development of components and promoting the development of completely new approaches. Since all of this as well as the experiences with structural

Table 5.1 Factors driving the development of passive and low energy buildings

	Reduced heating costs	Comfort	Subsidies	Regulation
Supplier				
Architects	**75%**	20%	37%	25%
Prefab building manufacturers	55%	20%	**62%**	50%
Vent. system planners	55%	30%	**67%**	62%
Vent. system manufacturers	45%	20%	**80%**	75%
Demand				
Property developer	**62%**	22%	**62%**	50%

Source: Bühring et al., 2004.

optimization are also used in EnEV new buildings and building refurbishments, the passive house approach is taking on an important role as technology driver which should not be underestimated.

Only technologies which support the learning of both system integrators and tradespeople address the bottlenecks in the diffusion of passive buildings. These technologies can be related to components of the passive house system which are easier to integrate, such as new types of wall systems which facilitate the construction of an air-tight building envelope. These technologies can also be related to the development of new planning, coordination and construction tools.

Nevertheless, the further development of components bolsters the market introduction of passive houses, above all when renovating older buildings with their additional technical challenges. The main components to be cited include

- Vacuum insulation panels, which achieve the same degree of thermal insulation as conventional insulating materials but at reduced wall thickness and so open up new applications when renovating old buildings, as well as improving architectural creative freedom and reducing the space required (benefiting the size of the building plot required or that of the internal rooms).
- Vacuum glazing, which offers greater possibilities for architectural design just like vacuum insulation panels due to its reduced thickness and lower weight compared to triple glazing of similar quality.
- The window frames which are currently, unnecessarily, still one of the thermal weak points among the passive house components.

Ventilation systems as active components of passive houses still have an optimisation potential with regard to energy demand, degree of heat recovery, smells, noise, installation optimisation etc.

We also emphasized that standards play a crucial role in reducing the uncertainty both on the supply and the demand side of the passive house market. Policy intervention should ensure the viability of the established standard and its protagonists. Also having discussed the central role of passive house related information diffusion networks and their role in matching supply and demand and reducing search costs on the market, the appropriate infrastructure should be supported to sustainably foster the generation, growth and persistence of the network. Targeted educational and information programmes for actors on all levels and at each stage of the passive house value chain are required to reduce the investment on the side of the passive house suppliers prior to entering the passive house market.

References

Bühring, A., Leuchtner, J., Krug, P., & Schüle, R. (2004). *Marktpotenzial für Passivhäuser und 3-Liter-Häuser*. Freiburg: Fraunhofer ISE,.

Bühring, A., & Russ, C. (2000). Lüftungs-Kompaktgeräte: Angebot, Erfahrungen und Weiterentwicklung, 4. Passivhaustagung

Feist, W. (2001). Energieeffizienz, In Rebhan (Hrsg.) "Handbuch Energie", Heidelberg.

Feist, W., Peper, S., & Pfluger, R. (2004). Ein nordorientiertes Passivhaus – Endbericht -, PHI, Darmstadt.

Haum, R., & Nill, J. (2004). Zeitstrategien ökologischer Innovationspolitik bei Wohngebäuden. Berlin: BMBF.

Horn, G. (2005). Mit dem Passivhaus den Markt erobern – Gutes Produkt für schlechte Zeiten. In Feist, W. (Ed.), Zukunft aktiv gestalten, Tagungsband, 9. Internationale Passivhaustagung.

Jakob, U., & Eicker, U. (2005). Betrieb und Simulation von Diffusions-Absorptionskältemaschinen zur solaren Kühlung, KI 11/2005

KfW-Förderprogramm (2005). Ökologisch Bauen. Förderprogramm Nr. 144, 145. 07/2005. KfW, Frankfurt.

Kleinhempel, A.-K. (2005). *Innovative Dämmstoffe im Bauwesen, Forschungsstand und Marktübersicht*. Bremer: Bremer Energie Institut.

Klingendberg, K. (2005). The Ecological Construction Laboratory: Strategies for Passive House Implementation in the US. In Feist, W. (Ed.), *Zukunft aktiv gestalten, Tagungsband*, 573–578. 9. Internationale Passivhaustagung.

Oberhuber, A., Amann, W., & Bauernfein, S. (2005). *Benchmarking Nachhaltigkeit in der Wohnbauförderung der Bundesländer, Berichte aus Energie und Umweltforschung*. Wien: BMVIT.

Plöderl, H. (2005). Architektur und Passivhaus. In Feist, W. (Ed.), *Zukunft aktiv gestalten, Tagungsband*. 9. Internationale Passivhaustagung.

Pöschk, J., & Heimann, C. (2004). *Bauen für die Zukunft*. dena.

Rasmussen, D., & Grocholl, M. (2005). Neue Wege für das Passivhaus-Marketing, In W. Feist (Ed.), *Zukunft aktiv gestalten, Tagungsband*, 429–4349. Internationale Passivhaustagung.

Schneider, N. (2003). Anwendungen nanoporöser Dämmstoffe im Bauwesen, Fachtagung VIP-Bau.

Schnieders, J., & Hermelink, A. (2006). CEPHEUS results: Measurements and occupants' satisfaction provide evidence for Passive Houses being an option for sustainable building. *Energy Policy, 34*, 151–171.

Schwab, H. et al. (2003). Entwicklung und Anwendung von evakuierten höchsteffizienten Dämmungen für Gebäude, ZAE Bayern.

Siraki, K., & Wagner, H.-J. (2006). Primärenergiebedarf von Gebäuden mit hohem Energiestandard, LEE Uni Bochum.

Wall, M. (2005). Terrace houses in Gothenburg – the first passive houses in Sweden. In W. Feist (Ed.), *Zukunft aktiv gestalten, Tagungsband*, 561–566. 9. Internationale Passivhaustagung.

Chapter 6
Carbon Dioxide-Free Power Stations/Carbon Dioxide Capture and Storage

6.1 Description and Significance for Energy Supply

In 2003, global emissions from fossil fuel use amounted to 25 GtCO_2 (IEA, 2005). More than 50% of this was attributed to large stationary emission sources (> 0.1 MTCO_2 yr.), i.e. these sources are the prime candidates for CO_2 capture (IPCC, 2005). The bulk of these consist of power plants along with a much lower number of installations from manufacturing and from the production of transportation fuels (see Table 6.1).

Estimates of future carbon dioxide emissions vary of course, but by 2030 they could be as much as 60–90% higher than the Kyoto benchmark value of 1990, depending on the measures taken to control them.

Within Europe, Germany is the largest carbon emitter, responsible for 25% of the CO_2 emissions of Annex II Countries in Europe (IEA, 2005). Electricity generation accounts for a large share of these emissions since around 57% is based on coal, lignite or natural gas (VDEW, 2006, Arbeitsgemeinschaft Energiebilanzen, 2005; see Table 6.2).

The existing company structure in the electricity industry faces the liberalisation of the European electricity and gas markets together with increasing trends towards decentralised energy conversion technologies. Furthermore, the start-up of the European Emissions Trading System (ETS) is strongly influencing framework conditions for investment decisions.

Table 6.1 Global profile of large stationary CO_2 sources by type of industrial activity (>0.1 Mt CO_2 per year), 2004

Fossil fuels	Number of sources	Emissions in Mt CO_2year
Power	4,942	10,539
Cement production	1,175	932
Refineries	638	798
Iron and steel industry	269	646
Petrochemical industry	470	379
Others	n.a.	83

Source: IPCC, 2005

E. Jochem (ed.), *Improving the Efficiency of R&D and the Market Diffusion of Energy Technologies*, DOI 10.1007/978-3-7908-2154-3_6, © Physica-Verlag Heidelberg 2009

Table 6.2 Shares of energy carriers in electricity generation, Germany 2004 and 2005

Energy carrier	2004	2005
Nuclear energy	30%	29%
Lignite	27%	26%
Hard coal	22%	21%
Renewable sources	10%	11%
Natural gas	8%	10%
Other	3%	3%

Source: VDEW, 2006

Like the other EU-15 Member States, Germany has adopted an emissions reduction target of 8% under the Kyoto Protocol. In the so-called Burden-Sharing Agreement of EU Member States, Germany pledged to reduce its emissions by 21% within the first commitment period (2008–2012) compared to 1990 levels. The targets under discussion for the post-Kyoto periods are even more ambitious, e.g. 40% CO_2 emission reduction by 2025.

Many actors in the global energy system believe a change in electricity production is necessary in order to mitigate climate change. The main challenge to the electricity sector in Germany, as it is worldwide, is how to avoid the CO_2 emissions from the use of fossil fuels. The solutions proposed range from fusion as a long-term option through increased nuclear fission electricity generation and increased energy efficiency to electricity generation from renewable energies. Among the less revolutionary options, when looking at the entire energy system – it has been proposed to capture the CO_2 produced in energy conversion processes and store it outside the atmosphere. Such a procedure would make it possible to maintain existing patterns of energy conversion and energy use and thus retain the present structure of the energy system. As a consequence, this option permits an evolutionary step to be taken and – with high probability – it could be implemented relatively quickly compared to other options. At present, however, its technical and environmental feasibility are far from given, which affects its economic feasibility as well.

6.2 Micro-Level Analysis

6.2.1 Technological and Economic Challenges

CO_2 free power stations[50] are in a very early stage of development, and there are several fundamentally different concepts competing for the development to market readiness and it is not yet clear which of these concepts will prevail. For this reason,

[50] Although fossil fuelled power stations with CO_2 capture will not be entirely "CO_2-free", the term "CO_2 free" is used in order to differentiate from the term "low CO_2 power station" which is often used in Germany with respect to highly efficient power stations without CO_2 capture.

CO_2 free power stations as a whole were chosen as the subject for analysis including the step of storing CO_2. An existing solution for storage is a prerequisite for the realisation of CO_2 free power stations and hence an essential part of the innovation system.

Achieving the so-called decarbonisation of fossil fuel fired power stations involves capturing the CO_2 at some stage within the energy conversion process for which different technology concepts are presently being discussed. **Post-combustion** concepts leave the conventional power conversion process generally unchanged and foresee an additional step in the flue gas cleanup in which the CO_2 is removed using a chemical solvent. **Pre-combustion** concepts aim to achieve the separation task before the main energy conversion. A synthesis gas is generated from the fossil fuel which has a low-carbon, energy-rich gas (usually hydrogen) and CO_2 as its main components. Usually by applying a physical solvent, the CO_2 is then separated before combusting the synthesis gas. In **oxyfuel** concepts, the separation is done at an even earlier stage of the entire process by splitting air into oxygen and nitrogen and then firing the fossil fuel with (almost) pure oxygen which produces a flue gas containing mainly CO_2 and steam, which can simply be condensed.

All of the concepts considered involve the compression of the separated CO_2 to its dense state (supercritical conditions) or possibly liquefaction. When the CO_2 has been changed into a transportable state by either of the two processes, it is sent to the storage site by either pipeline or ship. Other means of transport such as trucks or rail are not suitable for the anticipated amounts of CO_2. Geological storage will most likely take place in either oil or gas fields or in deep saline aquifers by injection of the CO_2 through injection boreholes.

The main challenges connected with CO_2 free power stations are not in the principal realisation of the concepts, which are more or less certain, but in achieving the technical and economic parameters required by the power generation industry. The first and largest challenge is the reduction of the specific energy requirement to capture the CO_2 within the energy conversion process. The energy requirements projected when applying current state-of-the-art technology would cause prohibitively high electricity generation costs and would also increase resource depletion. The second challenge is to design power station technologies with CO_2 capture that are at least as reliable as conventional power stations. A significantly lower availability of power stations with CO_2 capture would reduce the economic viability of such plants to a degree that is not acceptable for electricity producers. The third challenge for R&D lies in the development of storage solutions for CO_2 and proof of their long-term safety.

6.2.2 Status of CCS in the Technology Cycle

The development and market entrance of large-scale power generation technologies can generally be described as a chain of upscaling processes before actual "market diffusion" takes place. Market diffusion can be understood here as the large-scale construction and sales of commercial units of the respective power station technology. The upscaling processes undertaken before commercialisation result from a strategy that minimises the financial risks inherent to technology development. The first step when developing a new power conversion concept consists of the so-called bench-scale and lab-scale tests, where the components are built and operated in a laboratory with reactor diameters of several centimetres to some tens of centimetres. After successful lab-scale tests, the next step usually taken is the construction of a pilot plant which already has a thermal capacity of some megawatts or some tens of megawatts. The pilot plant allows the technology concepts to be tested in an industrial environment and under real-life conditions without the financial risk of a full scale plant.

The last step before complete commercial availability of the technology comprises the construction of a full-scale or almost full-scale demonstration plant. With respect to the power generation technologies discussed here, this means that a demonstration plant would have the scale of several hundreds of megawatts. In such a plant, the technology concept being tested is fully employed. However, in order to still minimise the financial risks involved, the plant may be constructed e.g. with not all the features for long-term maintenance included. The plant would thus have a reduced technical lifespan. It should be noted that this description refers to an idealised process of upscaling which does not necessarily always take place in such a clear way.

Until spring 2005, the research and development work on CO_2 free power stations had taken place on a lab-scale and it was anticipated that this state would continue for some years. At that time, the technology was clearly still in phase 2 of the technology cycle: due to the fact that this comprises a systemic innovation, most inventions had been made already and fundamental research questions answered. A growing community of researchers were working on implementing and upscaling the technology. This phase is termed "euphoria", but should not be misunderstood as implying that the actors in the innovation system do not perceive the potential risks and pitfalls of the technologies concerned and of the societal framework for CO_2 free power stations. In the interviews, activities were reported to be constantly increasing while maintaining a diversity of approaches and industrial players inside and outside Germany were accelerating the pace of development by announcing upscaling projects to approach industrial dimensions.

In the summer of 2005, however, the Swedish utility, Vattenfall, announced that it would start construction of a pilot plant in Germany in 2006. Further, the oil and gas producer, BP, announced plans to construct a CO_2 free power station at Peterhead in Scotland, which aims to start operation in 2009. During spring 2006, the utilities RWE and E.ON also announced power plant projects implementing CO_2 capture. These announcements are the first indications of a real market for CO_2 free power generation technologies. However, the nature of the projects concerned indicates that we are still really dealing with prototype technologies here. As all of the announced projects carry a reasonable risk of failure, which could lead to disillusionment, CCS should be viewed as still in phase 2 of the technology cycle.

In line with the general analysis of the state of market development, most of the interviewed actors reported there were no market studies available. One actor mentioned that they were using studies elaborated by consultancies active in the energy business on their behalf. As well as the lack of specific market studies, so far there has been no use of brand names for technologies in the field of carbon dioxide free power stations either, but actors also reported that such branding is generally unimportant with respect to power generation technologies.

From a methodological point of view, it should be possible to provide concrete information about the position of a technology in the technology cycle based on analyses of quantitative indicators such as patent intensities. However, performing the necessary patent analysis for CCS technologies harbours particular problems.

First of all, the technology routes discussed for CO_2 capture do not involve a radical change in power conversion technology concepts. Instead, CO_2 capture usually means employing conventional technologies which have been further developed together with additional components that are already in use in other industries such as air separation units or scrubbers which are already state-of-the-art applications in process engineering. Thus, the research and development work on capture technologies and the resulting patents are mixed to a certain degree with other more conventional development efforts and the patents generated from them.

This effect can be shown based on the example of IGCC technologies. Gasifiers, for example, which are a core element of IGCC, were developed several decades ago. A lot of research and development work was carried out in the 1980s and early 1990s and resulted in the realisation of a few plants such as the one in Puertollano or in Buggenum. Developing this technology was mainly motivated by the desire to improve the overall efficiency of power plants using coal by applying a combined cycle. Further, the idea was to reduce air pollutants more efficiently with the IGCC approach. At the time the technology was developed and the first plants realised, CO_2 capture was not an issue at the level of technology development. Patents related to gasifiers and IGCC plants in this period are therefore not an indicator for activity in the field of CO_2 free power plants. Since 2000, however, it can be assumed that a large share and probably even most of the patents on IGCCs are related to pre-combustion capture technology. Nevertheless, it is virtually impossible to determine with the necessary certitude when and over what period of time the focus of IGCC-related patents changed.

Secondly, many other technology patents are categorised within the patent classes containing CCS technologies. In order to produce useable statistics on CCS-relevant patents, these non-relevant patents have to be filtered out. The commonly applied approach for this task – using key words within patent classes – proved not to work for CO_2 capture technologies as a large number of patent abstracts do not contain relevant key words. Instead, in many cases, the relevance of patents for CO_2 capture technologies was only able to be based on the content of the abstract.

The problem of using key words in relation to the energy conversion processes can also be illustrated with the example of patents related to gas turbine processes for electricity generation. Patents of such processes are related to CO_2 capture if operated with a carbon free fuel. However, the patents do not usually contain an expression like "carbon free fuel". Using just the key words such as "carbon" or "hydrogen" (which is the fuel of gas turbines in many pre-combustion concepts) will not give usable results as it is the context of these key words which determines whether the described process is intended to avoid CO_2 emissions. Analysing the abstracts in this field also provided evidence once again for the attempts made by patent applicants to hide the ultimate purpose of their development. This problem was encountered for example for the membrane reactors described in patents. Due to the very general nature of the description, it was not possible to decide whether such a patent is relevant for CCS or not, although there was a good probability that this was the case.

An analysis of the abstracts was done for the technology of oxygen separation by membranes. This technology might reduce the energy penalty of oxyfuel processes by a significant amount. A database query in the relevant patent class for gas separation by diffusion combined with the key word "oxygen" produced about 100 patents up to the year 2002. Of these, roughly 30 had to be excluded as not relevant for CO_2 capture and storage. These included patents describing technologies for nitrogen production, for medical applications, for scuba diving enriched air production or for oxygen enrichment in the combustion air of mobile combustion engines.

Besides their nature as a systemic innovation, CO_2 free power stations are not a single technology but involve different technology concepts, each of which in turn consists of various technology components. As a consequence, the range of patent classes in question is very broad and so far it has not been possible to identify an exhaustive set of patent classes. Ultimately, the amount of work necessary to produce meaningful results proved to be too large for the resources that could be allocated to this task.

In the area of publications in scientific journals, the problems were not as large as those encountered in patents. In journal articles, authors do not usually conceal the purpose or the overriding objective of their work so that key word searches could be applied. The results of the publication analysis are shown in Fig. 6.1. A peak in publications in the second half of the 1990s can be observed followed by a small decline around the turn of the century. In theory, this shape could be interpreted as an example of the technology cycle. However, other explanations are called for here given the low number of publications identified and the fact that no phase of disillusionment was reported by the interviewed actors. The shape of the curve is possibly influenced by the PACE research programmes like the EU-Framework programmes, which have a significant influence on research and subsequent publication activities.

Such articles do not fully represent the publication activity in the field of CO_2 capture and storage. There are several series of conferences where the scientific community of this field discusses the research and development results. The Greenhouse Gas Control Technologies Conferences with the IEA-Greenhouse Gas Research and Development Programme acting as guardian are probably the most important. These conferences are receiving more and more attention and an increasing number of submissions of papers and presentations. The conference proceedings, although some of them are printed by scientific publishers, were not covered in the literature databases. As a result, an important body of scientific literature was

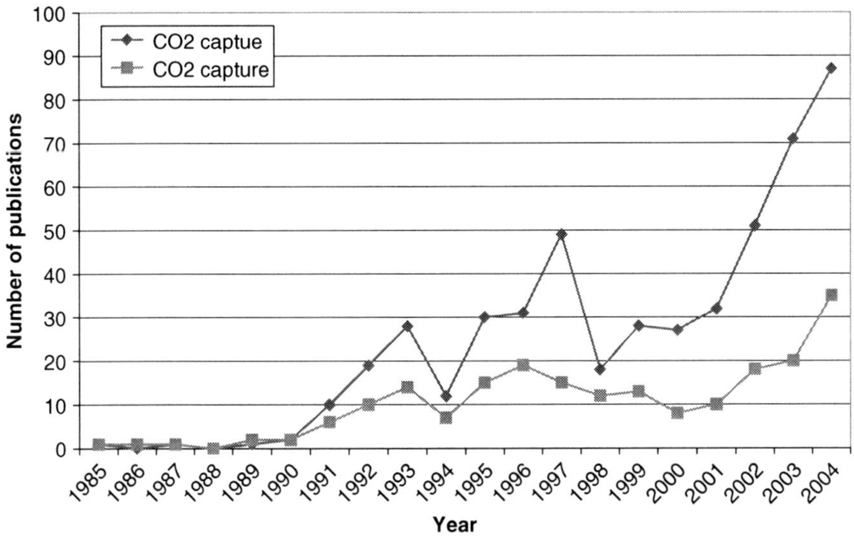

Fig. 6.1 Trend in scientific journals of publications on CO_2 capture and CO_2 storage, 1985–2004

not included here. The study by Radgen et al. (2005, to be published), which analysed the most important conferences from scratch, evaluating each article based on the printed proceedings, shows that the trend of publication activity found there is quite similar to the one found for journal articles here. A comparable small decline in publications at the very end of the 1990s could be observed there within the overall rising trend. The absolute number of publications found in Radgen's study is about three times as many.

From the interviews it was again very clear that the market for CO_2 free power station technologies in Germany is solely dependent on the development of climate change mitigation policies which assign costs to the emissions of CO_2 to the atmosphere. There are hardly any other co-benefits of CO_2 free power plants compared to conventional ones apart from the fact that they produce only a fraction of the CO_2 emissions. In the framework of European gas and oil production, there might be some extra value generated from the sale of CO_2 for enhanced oil recovery operations. This is not of primary interest to the German electricity industry because of the long transport distances to the North Sea oil fields.

Especially actors from industry stressed that a clear long-term perspective about the reduction requirements of greenhouse gas emissions is necessary for the market entrance of CO_2 free power stations. In contrast, one actor described the situation as more a process of gaining experience with the policies for mitigation of climate change, especially with the European Emissions Trading System. Through this process – probably after some years in the post-Kyoto period – sufficient confidence might have accumulated to take investment decisions about CO_2 capture and storage technologies without the often demanded "long-term setting of framework conditions".

6.2.2.1 Breakthrough During the Last Five Years

Over the last five years (or less), the opinion of most relevant actors in Germany towards carbon dioxide free power stations and CCS has changed dramatically. Prior to this change, one actor reported that he had been "laughed at" when suggesting CCS to a power supply company and another reported very strong resistance in the power supply industry to the idea of carbon dioxide free power stations in 2002. Even in 2005, some actors did not believe the European Emissions Trading System (ETS) would be implemented. By the end of 2005, however, it was obvious to all actors that CCS and carbon dioxide free power stations are serious options, even though not all of them agree that these are or will be the best available way to reduce carbon dioxide emissions. How did this very rapid change in attitude come about? Various factors were mentioned by our interview partners, and many individual theories abound.

The electricity industry, which has been characterised by interview partners as very conservative, reacted very late to the introduction of the European Emissions Trading System, seemingly waiting for the ETS to materialise and to be given proof that putting a price on CO_2 emissions would really be implemented. It was claimed that technology producers need to be "ahead" of the market in the sense that they have to develop technological products and have them ready for when the power generation companies decide to purchase them.

The public research programmes in Germany have only adapted their scope to include CO_2 capture and storage technologies quite recently. Since 2003, project applications have been accepted by the administrative bodies for the federal research programmes and actual projects with scope for CO_2 capture have been receiving support since 2004. This is quite late compared to the international framework, on a European level as well, where large projects have been supported since the late 1990s.

6.2.2.2 Next Change: Will Security of Supply be Dominant?

Actors hazard different guesses about the time span necessary to develop and construct a "reasonable" carbon dioxide free power station, which range from eight to twenty-five years for the first demonstration plant in Europe – be this in Germany or elsewhere. One interviewee defined "reasonable" as a plant with less than 10% loss of energy efficiency compared with conventional power plants and less than double the investment cost. Most actors do not expect complete market penetration. Instead, they expect single carbon dioxide free power stations by 2025 and possibly a respectable share of the market later on. There was also, however, one interview partner, who was quite confident that all large-scale fossil fuelled power plants will be equipped with CO_2 capture technologies in the future, as long as the climate policies do not deviate significantly from the path envisaged today. Overall, it has to be concluded from the interviews that it is still uncertain whether carbon dioxide free power stations will be built at all in Germany. The actors in this field are very aware of this uncertainty: they handle the subject with care and pay great attention to developments in the area, but do not expect final decisions about the implementation of this technology before 2015. New basic research on other innovative technologies might be necessary which would slow down the process significantly. Additionally, there will need to be a considerable amount of time invested in research on storage because this has to include evidence of long-term storage integrity. Nevertheless, actors in the subsystem of storage expect "their" part of CCS to be ready to implement first.

All interview partners agree on the necessity to considerably reduce carbon dioxide emissions in order to mitigate climate change. Therefore, the interest in carbon dioxide free power stations is not expected to die away soon. But actors do see alternatives to this route, especially nuclear power stations and a shift from coal to gas (in the short run) and to renewable energies (in the very long run).

The loss of energy efficiency associated with the capture of CO_2 is the main technical bottleneck. Taking into account that securing energy supply will probably become more difficult, some actors warn that this may hinder the development of carbon dioxide free power stations because of their greater demand for fossil fuels. CCS might be useful to mitigate climate change, but will tend to aggravate energy supply problems.

In Germany, the targets of the Kyoto protocol will probably be met by making conventional power plants more efficient and also by profiting from structural changes in the 5 Bundesländer that made up former East Germany, thereby saving fossil fuels and reducing carbon dioxide emissions. But these reductions will not be sufficient to meet the more ambitious targets planned for the post-Kyoto periods – if fossil power plants are to remain an important part of energy supply, they will need to be virtually carbon dioxide free. In particular, if political actors decide to enforce the ambitious target to reduce carbon dioxide emissions by 40% by 2025, carbon dioxide free power stations will be unavoidable.

6.3 The Innovation System of CCS

6.3.1 The Overall System

To build CO_2 free power stations, technologies are needed for (1) capture, (2) transport and (3) storage of carbon dioxide. The innovation systems for these three tasks in the process chain of CO_2 capture and storage are closely interlinked and form the overall innovation system of CCS. From an innovation-centred point of view, the systems of capture and storage are more challenging than the system concerning transport, because the transportation of carbon dioxide is seen as technically and probably also politically less problematic. Since several of the leading players are active in research concerning both the capture and the storage of carbon dioxide, both systems will be described and analysed as one, unless greater differentiation is required. The question concerning which of the three major technology routes for carbon dioxide capture will predominate is still open. Although some actors are concentrating on one particular technology, no one is completely neglecting the other two options: the innovation system of carbon dioxide capture and storage has not yet (but may in the future) split into more specific technological innovation systems. It is worth mentioning that some actors pointed to a possible parallel between the introduction of desulphurisation technologies in the 1980s and the anticipated introduction of CO_2 capture technologies. According to them, the initially applied desulphurisation technologies have long been displaced by more efficient and less costly technologies. With respect to CO_2 capture technologies this means that significantly modified concepts or even entirely different concepts than those discussed today may dominate the technology market in the long run.

The main driving forces of the innovation system in Germany are of a political nature. Carbon dioxide free power stations as such are energetically less efficient compared to highly efficient power stations without carbon dioxide capture. They are economically less efficient as well, unless the costs for CO_2 emissions are internalised. Therefore, carbon dioxide free power stations will only make economic sense to utilities under a regime where CO_2 emissions are associated with costs due to the money then able to be saved from reduced carbon dioxide emissions.

The innovation system for CCS is centred on a technology that does not yet exist, at least not at the scale ultimately required, i.e. suitable for large-scale fossil fuelled power stations. Therefore, there is no current demand, only potential demand. The only exception to this in Germany concerns Vattenfall's planned pilot power station "Schwarze Pumpe". Presumably from 2008 onwards, the oxyfuel technology will be used here to capture the carbon dioxide from the generation of the equivalent of 20–30 MW of electric power – which is still far below the ultimately targeted size (1000 MW for normal fossil power stations). At this time, this pilot plant will be the first (near-) industrial scale plant where electricity is generated from solid fossil fuels with CO_2 capture.

The innovation system for CCS in Germany consists of a limited number of important actors and other more minor ones.[51] Without actual demand and supply, the market subsystem must be analyzed as a *potential* market, consisting of four big potential *demand* companies in Germany, RWE, E.ON, Vattenfall and EnBW, and two suppliers of conventional power station technologies as the potential *suppliers* of integrated power plant technologies with CO_2 capture, Siemens and Alstom. Of course, international competitors might also enter the market but so far, these are not prominently active in Germany. However, some actors stated that they see the power conversion technology market as being completely international in nature and that foreign suppliers market their products in the same way as domestic ones. Additionally, some interview partners presume that international oil and gas companies will enter the German market to build their own power stations. This could substantially alter the technology market as well as the electricity market.

---

[51] It should be kept in mind that the development and manufacture of power generation technologies is an international business and that the development of CCS technologies also takes place in an international framework. Nevertheless, the scope of Germany was chosen for this analysis in order to investigate the chances and opportunities resulting from domestically based activities.

Unlike the clearly structured techno-economic system, the research and the political systems need be looked at in more detail. The responsibility for public research funding of technologies to reduce carbon dioxide is distributed among different ministries, causing some confusion due to their different respective strategic objectives, and research is spread over a number of institutions, ranging from the Federal Ministry of Economics and Technology, which is responsible for research in energy conversion technologies (with the exemption of technologies for the use of renewable energies), to the Federal Ministry of Education and Research, responsible for the natural science and mining technology research connected to CO_2 storage. Apart from these, the Federal Ministry for the Environment also conducts investigative research work on CCS motivated by the possible future competition between renewable energy technologies and CCS and by the possible need to regulate storage under water and waste legislation.

6.3.2 Components of the Innovation System

6.3.2.1 Demand: The Power Supply Industry

Within Germany, the demand for large power station technologies is limited to the existing power supply industry with perhaps some independent power producers. Some of the municipal utilities operate electric power stations, but usually only the "big four" electric power companies – RWE, Vattenfall, E.ON and EnBW –, which emerged from the former regional monopolies, are said to be able to build and operate a plant with CO_2 capture and storage. Contrary to this more commonly expressed view, one interview partner also mentioned that municipal utilities might be able to build and operate such a plant as part of a local or regional sustainability strategy. Generally one can assume that larger utilities have a better chance to internally hedge the risk inherent in building a capital intensive power plant based on a new technology such as carbon capture. From a European perspective, a number of other electric power supply companies may demand carbon dioxide free power stations, but this does not change the character of the system which is often described as low in competition. All interview partners agreed that in Germany the power supply industry is the leading actor in the field of CO_2 free power stations.

With its extremely narrow field of potential buyers, the innovation system for carbon capture and storage constitutes a special case among the technologies investigated. Instead of relying on market forecasts based on data that can be analysed statistically, here the individual strategies of the large players involved are of pivotal importance.

It is not unlikely that demand for carbon dioxide free power stations may *only* come from the four named companies. At present, the power supply industry is only doing research on carbon dioxide free power stations to the tune of several million Euro per year. The only exception and therefore the only *current* demand for carbon dioxide free power station technology is from Vattenfall who are building a pilot plant at "Schwarze Pumpe". This pilot plant should actually be interpreted as part of the research efforts, too, because many technological details are still unresolved. Vattenfall is determined to finish construction of this plant by 2008, maybe even by the end of 2007. Its motivation to build such a cost intensive ($\approx$70 million Euro) pilot station cannot be the direct economic benefit, but is of a more strategic nature. It may give Vattenfall the technological leadership with respect to oxyfuel technologies or even with respect to CCS as a whole and may improve its reputation as an environmentally-friendly company.

Vattenfall is not ordering the entire system from one company, but is constructing the power station itself, probably in order to learn as much as possible and, as some actors assume, to involve a number of companies in building such power stations or parts thereof, thereby avoiding dependency on a single technology manufacturer. It is expected that all large power conversion technology suppliers like Siemens and Alstom will be providing components for the pilot plant employing oxyfuel technology.

Unlike Vattenfall, RWE used to be reluctant about the idea of carbon dioxide free power stations and concentrated instead on improving the efficiency of conventional power stations. RWE does participate in some of the large European research projects such as "CO_2 Sink" or ENCAP, but is not involved in the Dynamis project which is judged as being a key project for the development of CCS by the European Commission.

E.ON is seen by some actors as also concentrating its research on efficiency improvements of conventional plants rather than carbon dioxide free power stations. However, as E.ON Ruhrgas AG is part of the company, there is an interest in the field of transporting and storing carbon dioxide. There is no reason to assume that E.ON is not aware of the idea of CCS, but in contrast to other utilities, its strategists are probably not convinced that this technology will ultimately be implemented.

Among the large utilities in Germany, EnBW is less active in this field as it is relatively small with an even lower share of fossil power stations in its current portfolio. Furthermore, the geographical location of this company's plants is not very favourable for carbon capture and it seems difficult for large utilities to find suitable sites outside the area of their former regional monopoly.

One interview partner characterised the situation in Germany as follows: of the large German utilities, the greater the degree of public ownership and the larger the number of lignite-fired power stations operated, the more interested it is in carbon dioxide free power stations. This interest is motivated by public relations on the one hand and by the high specific carbon dioxide emissions of lignite-fired power stations on the other which make CO_2 capture particularly promising. All companies emphasized that they are missing the clear, long-term framework conditions which would allow them to plan and build power stations with or without carbon dioxide sequestration. While some interview partners assume that power supply companies are waiting for these conditions to apply and trying to prolong the use of their older power stations in the meantime, others report that, on the contrary, power supply companies are trying to build conventional power stations as quickly as possible to forestall any future obligations of substantial emission reductions.

Internationally, a demand for carbon dioxide free power stations may also arise from the oil and gas industry caused by the demand for carbon dioxide for enhanced oil recovery (see below).

Other smaller utilities like the municipal utilities are not expected to become important players in the innovation system. Most of our interview partners believe the same is true for independent power producers (IPP), which are more common in the Netherlands and the UK. However, some actors take the rise of IPPs into account, especially if the big power supply companies delay too long in building carbon dioxide free power stations.

6.3.2.2 Manufacturers

Large-scale power conversion technologies are supplied by international companies. There is no truly domestic manufacturer. This is due to the fact that the production of these energy conversion technologies is characterised by very high fixed costs arising from the need to generate scientific and technological skills and patents, keep a highly qualified workforce and develop highly integrated technological products. All of this together can only be maintained on a high technological level by large international companies. In Germany, there are two companies with significant R&D and production – Siemens and Alstom – who could possibly supply complete turnkey carbon dioxide free power stations. In order to do so, they would need to integrate components from other sub-suppliers. Of these two companies, Siemens' head office is in Germany and a substantial share of its R&D takes place here. Alstom's head office is in France, but there are R&D activities and production capacities located in Germany. The sub-suppliers required by both are also available in Germany with a few exceptions. Linde and Lurgi were mentioned most often in the interviews as component suppliers. Some international competitors are reported as able to enter the German market, especially General Electric or Mitsubishi Heavy Industries. With respect to boilers, Babcock-Hitachi-Europe also plays a role.

Fierce competition is not yet very common in the innovation system of carbon dioxide free power stations because there is no real demand and therefore no money to earn at the moment – the interviewed actors are not even sure which technology will prevail, i.e. will actually be able to be sold. Furthermore, research is still often at the level of basic research, and cooperation here between companies is common and widespread in order to share costs. However, the statement was also made that the challenges and work needed could shift very quickly from fundamental research activities to actual technology developments.

The research and development strategies of both large technology suppliers (Siemens and Alstom) have been characterised by the actors as open regarding the main technology routes of pre-combustion, post-combustion or oxyfuel. Nevertheless, Siemens and Alstom seem to follow somewhat different (and competing) strategies stemming from the respective company portfolios. It can be concluded that neither of the technology suppliers has a set opinion about which technology route will be successful on the market. Therefore both try to keep abreast of the developments in all the major routes. No indications could be found that either of them intends to alter their present technology portfolio (i.e. acquiring a component provider) in order to increase future competitiveness with respect to CO_2 capture technology.

Research in CCS is mainly driven by the expected demand from the power supply industry, but technology manufacturers also try to stay "a step ahead" to balance the time they need for research. Additionally, being involved in any overtly environmentally-friendly technology can be an asset for public relations.

For all the relevant manufacturing companies, CCS technologies themselves will only generate a smaller part of their turnover so that they will not necessarily be directly dependent on them. However, the actors imagine that at least a substantial share of the market for power generation technologies could depend on CCS. So there is a direct need for the technology suppliers to make their technologies at least compatible with CCS technologies. According to the majority of interview partners from industry, nuclear power is one other way to achieve climate policy targets, and none of them forgot to mention that option in our interviews.

6.3.2.3 Oil and Gas Industry

Oil and gas companies, though not of German origin, are important actors in the international CCS innovation system and may become important in Germany, too. Some interview partners even suggested placing them at the centre of the innovation system because they are believed to have and to continue to accumulate experience in all the important parts of the process chain of CCS and will promote this for several reasons. First, these companies need large quantities of carbon dioxide for enhanced oil recovery (EOR) operations in the North Sea oil and gas fields. As a result they are very interested in the development and promotion of CO_2 capture technologies. Second, they want to continue to sell their products and fear carbon dioxide emission restrictions may affect sales, so they are interested in politically accepted – i.e. carbon dioxide free – fossil power stations. Third, they have wide experience with geotechnical operations and to some extent also with injecting CO_2 into oil and gas fields. Hence, they see the opportunity to expand their business activities based on this experience. Finally, they have enough financial clout to build CO_2 free power stations. Therefore, oil and gas companies may become competitors to German power supply companies in this field, or may intervene in the German innovation system in other ways. For example, some interview partners report that Shell is informally promoting and providing funds for CCS research projects.

6.3.2.4 Other Industry

For pipeline operators such as E.ON Ruhrgas, transporting CO_2 may become a new market activity. The transportation of gases is a comparatively low challenge for those companies with experience in transporting natural gas, and these companies sense a possible new market in transporting carbon dioxide. Since E.ON Ruhrgas is owned by E.ON, E.ON may have a strategic interest in combining the sequestration and transport of carbon dioxide. One interview partner mentioned that a legal obstacle still remains here since, so far, right of way is only granted for natural gas pipelines

6.3.2.5 Public Research

Although the innovation of fossil carbon dioxide free power stations is still at an early stage, German universities do not seem to be central actors in the innovation system. There are a number of universities conducting research and there are also some ambitious projects such as e.g. ADECOS or OXYCOAL. Both of these aim at conducting basic research and therefore receive substantial support from public research funding of the federal government. This became problematic when Vattenfall announced their intention to build the pilot plant in Schwarze Pumpe much earlier than previously anticipated. From a formal point of view, some of the planned research work then had to be reconsidered as non-basic research and to some extent was even seen as expendable, causing problems and the need for reorganization. Additionally, the public funding share for some work has been reduced due to this change in the framework conditions. It seems to be inconsistent if, on the one hand, public authorities want to accelerate the innovation process and, on the other, EU regulations force the public funding of research projects to be reduced if industry makes a big step forward. This is exacerbated if the activities of industry are of a more risky nature and leapfrog the step of fundamental research. Because university projects and even entire university institutes normally rely heavily on public project funding, such an unexpected reduction – as occurred due to the announcement of the pilot plant in Schwarze Pumpe – can cause major problems.

Some university institutes seem not to participate in international cooperation in the field of CCS or at all. Instead, these institutes seem to generate funds only from German industry partners and especially from German authorities. The same concentration on the domestic scene also seems to exist for the academic discussion held by these institutes. This orientation is astonishing as many interview partners underlined the international nature of the technology development, the markets and also the industry actors.

The interviews revealed that international collaboration, mandatory to obtain funding from the EU, is not always welcomed by university actors. For them, it means splitting EU funds among many partners and at the same time increasing transaction costs, while the potential advantages of international cooperation do not seem to be exploited. In the interview it was also stressed that regulations for projects funded by the EU forced university institutes to accept foreign academic partners who did not contribute substantially to the success of the project. The question was raised why international partners should have to be involved, if the most qualified partners can also be found domestically.

With respect to the appraisal of cooperation within EU projects, it was questioned to what extent German universities really are the leading institutions in the engineering sciences for power conversion technologies and also whether it would be fruitful from an official German standpoint to develop strategies for German engineering sciences to identify highly qualified international partners and create networks of excellence.

Of course, universities fulfil more functions than just conducting research in joint projects for the innovation system, they provide a supply of new scientists and engineers to other actors in the innovation system. Some actors expect a shortage of geological engineers due to the decline of such research during the last decades, and many actors report that the engineers needed to construct power plants are quite hard to find even now and that this situation is expected to get worse.

One actor mentioned the close cooperation between the Norwegian University of Science and Technology (NTNU) and the applied research institution SINTEF as a very positive example of how to organise academic research internationally.

6.3.2.6 Intermediaries

Intermediary organisations like VGB and VDEW, the organizations of the power supply industry, are involved in most of the bigger cooperations. Although they were considered generally important in the electricity industry, they were not reported to play an important role themselves in the innovation system.

6.3.2.7 Financial

The big power supply companies have the financial capability to build fossil-fuelled carbon dioxide free power stations if they consider them feasible. The necessary investments of course will be very high. Since consequently any failure of these power stations will mean great losses to investors, manufacturers and the power supply industry need to be convinced that such a system will run continuously. Some actors were also adamant that the risks associated with the initial demonstration plants should be carried by public authorities in one way or another.

6.3.2.8 Societal Factors and NGOs

As already discussed, the societal and political goal of mitigating or slowing down climate change is the main driving force for carbon dioxide free power stations. But precisely those societal groups active in environmental politics are also seen as a threat to these power stations because they may resist carbon dioxide storage. Some actors reported, for example, that discussions with Greenpeace had not yet concluded and that Greenpeace's policy was still subject to change, but all actors agree that it will be necessary to involve environmental non-governmental organisations (NGOs) in the innovation process as early as possible.

So far, it is not certain whether society as a whole and residents close to carbon dioxide repositories in particular will accept carbon dioxide storage. All actors agree that CCS will not be implemented without societal acceptance, though all our interview partners themselves consider carbon dioxide storage to be unobjectionable. Additionally, it is unknown whether the entire technology of carbon dioxide free power stations will be seen as environmentally friendly because of reduced emissions, or harmful because of the remaining emissions and the increased resource use. A study of Shackley/McLachlan/Gough (2005) indicates that public opinion is still very volatile depending on the framing:[52] if presented as a bridging technology until renewables take over, fossil carbon dioxide free power stations were accepted by the sample group interviewed by the authors of the study; if presented as an alternative to a quick use of renewables, acceptance was reported to be low.

6.3.2.9 Political Actors

Based on the responses of German actors in the field of CO_2 free power stations, the German Ministry of Economics (BMWi) is by far the most important player in the political part of the innovation system when it comes to research funding, though there are a number of other actors on different levels: the EU (DG RTD), the Federal Ministry for the Environment (BMU), the Federal Ministry of Education and Research, and the Bundesländer, especially North Rhine-Westphalia, with their own (relatively small) research funding. But the BMWi claims to have and in most cases actually does have the leadership when it comes to projects in the field of CO_2 free power stations in Germany. However, actors from the research administration stress that they do not support a particular technique, but only set more abstract goals like the reduction of carbon dioxide, and then leave it to the rest of the innovation system to decide which option would be best. Nevertheless the influence of politics should not be neglected as there are indications that the government prevented research funds being spent on CCS around the turn of the century due to the political decision to foster renewables.

Some interview partners mentioned room for improvement concerning the coordination and cooperation between the Ministry of Economics and the Ministry of Education and Research as well the Ministry for the Environment. Very complex innovations which involve the competencies of different ministries will only be successful if all the links are active. In the case of CO_2 free power stations, both carbon dioxide capture (research responsibility: BMWi) *and* storage (research responsibility: BMBF) must be achieved. Since the BMWi and the BMBF are responsible for the research in these fields, they need to act in concert to achieve the necessary long-term goals. However, actors in industry from the capture technology side do not care too much about such political quarrels, their contact institution for concrete research funding in the political system is the BMWi or the Project Management Organisation Jülich (PTJ), and they do not report any problems with the funding of research projects due to suboptimal coordination within the political system.

[52] It should be noted that the study was conducted in the United Kingdom with non-representative panels and that the results cannot be directly transferred as valid for Germany.

6.3.3 The Role of Public Research Funding

As indicated above, the innovation of carbon dioxide free power stations is still in the first phase. This conclusion can be drawn even without the availability of quantitative indicators at this stage of the project, since the interview partners report a steep rise in the amount and intensity of activities still without any real market for the technology. In addition, there are no reports of any previous phase of disillusionment with respect to this technology, which indicates that, from the viewpoint of an ideal technology cycle, CCS is still in the initial phase. This position in the technology cycle has implications for the role of public research funding. All actors agree that there would have been hardly any research on carbon dioxide free power stations without public funding in the kick-off phase a few years ago, and many interview partners expect that CCS research would die away even now if public funding ceased. It was pointed out that commercial players are not able to spend money on research on such a long-term basis (or at least so they claim). Even the pilot station in Schwarze Pumpe, which does not receive any direct subsidies, benefits indirectly from publicly sponsored research (e.g. the ADECOS-project).

It seems that public research funding plays an important role in holding the system together, especially by offering financial support for research on a technology at a time when its economic value is still unknown. This uncertainty originates to a large degree from the influence of (climate) politics which is hard to predict. So, in a way, research politics have to overcome the shortcomings of energy and climate politics. Secondly and maybe even more importantly, the German government and the EU provide the necessary "glue" between the components of the system by organizing networks and cooperations like CO_2net and the Technology Platform for Zero Emission Fossil Fuel Power Plants. The German COORETEC programme also helps to bring actors together.

Actors from the BMWi and PTJ have multiple contacts with all other actors and cultivate these both formally (e.g. via project applications, yearly workshops, etc.) and informally. They use them to learn about new technical developments as soon as possible, interpreting any "weak signals" from industry. All actors agree that members of the BMWi listen carefully to their suggestions and arguments, thereby enabling them to influence the objective of public research funding, and cite the process experienced at the start of the COORETEC-programme and the COORETEC advisory board as examples. However, it seems that the BMWi was also taken by surprise when Vattenfall announced their plan to build an oxyfuel pilot power plant at Schwarze Pumpe so quickly, and then had to adjust the funding of related projects, thereby causing significant transaction costs.

Conversely, industrial actors try to screen political developments, too, and report that shifts in research funding are sometimes a more reliable indicator of long-term political aims and trends than programme papers or politicians' speeches. This is another function of research funding: indicating future political actions. The decision to support research in fossil power stations only in combination with CCS in the 6th Framework Programme of the EU has been interpreted by industrial actors as a strong indicator that they need to take CCS into account in their long-term planning.

However important public funding might be, some actors find the necessary bureaucracy of application and reporting procedures very demanding compared with the relatively small fraction of costs covered (20–50% for industrial actors, depending on the degree to which the research is considered to be basic), and are therefore somewhat reluctant to apply for funding. This cut-off is imposed by EU regulations to ensure that windfall profits are small. Not surprisingly, industrial actors recommend the American way of funding up to 100% of the costs, even for industrial projects.

Some actors stress the significance of public research funding for the recruitment and qualification of young scientists and engineers.

The BMU may be the responsible political actor for the storage of carbon dioxide. The responsibility depends on how storage will be regulated – under which area of legislation – mining law, waste law or water law. The BMU has responsibility for the latter two, whereas the BMWi is the relevant authority for the mining law. It should be recalled that, generally, the BMU has followed a policy of fostering renewable energies. However, a certain change in the policy towards CCS seems to be indicated from 2004. According to some actors, the BMU may now take CO_2 free power stations into account as a bridging technology until renewables are available on a large scale. In addition to the geological surveys in other European countries, another important actor in the field of carbon dioxide storage is the Bundesanstalt für Geowissenschaften und Rohstoffe (Federal Institute for Geosciences and Natural Resources (BGR)), which offers geological services in Germany to some degree.

EU funds are used for actual research projects such as large volume integrated projects, e.g. CASTOR, ENCAP, but also networking activities such as CO_2net or CO_2Geonet.

6.3.4 Regulation

Unlike the other three technologies discussed in this paper, regulation is of fundamental importance for the innovation system of CO_2 free power stations. Since they are less efficient both in terms of economics and energy than conventional fossil power stations, from today's perspective they will only be developed and built if new regulations limit emissions of CO_2 to the atmosphere. The regulation may consist of more indirect measures such as the currently used cap-and-trade mechanisms for carbon dioxide emissions, or, less probably, direct regulation requiring the reduction of carbon dioxide emissions from power stations. The trading of carbon dioxide emission certificates is still a very young system in Europe and associated with many uncertainties. For example, the price for emitting carbon dioxide is highly dependent on the quantities supplied and demanded – "ten million[53] tons can really make a large difference there" – as expressed by one interview partner.

[53]Compared to the total amount of emission allowances of roughly 2 billion tons of CO_2, 10 million tons are a very minor amount.

While the German federal level is the most important for research funding, with regard to regulations, all actors consider international conventions and European regulations to be more important than the domestic political actors who have to implement those regulations at the national level. Most important, of course, are the post-Kyoto conventions and regulations, which will set the required level of carbon dioxide emissions, and decide whether it is acceptable to reduce emissions by storing carbon dioxide underground. The second question is how the EU will break down these conventions into intra-Union regulations, i.e. whether there will still be a regime of carbon dioxide certificates and if so at what price? At the moment, emitting one ton of carbon dioxide is worth about 20 Euro, but this price would need to be about fifty to eighty percent higher in the long term to make CCS economically efficient, assuming technological improvements over the next 20 years.

In the national context, regulation is relevant for storage. Will carbon dioxide be defined as waste, a resource, or something sui generis? Depending on that definition, different regulations will apply to the storage of carbon dioxide, and different ministries will be responsible.

6.3.5 Cooperation and Interaction in the Innovation System

The innovation system for CO_2 free power stations and CCS is characterized by many networks and cooperative projects. Many of these projects and networks involve a large number of members from different groups, i.e. public research, power supply industry, manufacturers and other organisations. Some big networks do not involve the administration of funds, but are meant to foster communication between the actors. Our interview partners thought these networks fulfilled this task very well, helping actors to find their place in the innovation system, identifying promising paths, and, most importantly, setting up joint projects, although some actors are disappointed about the small yield of measurable benefits from these networks (COORETEC advisory council in Germany, Technology Platform and CO_2Net in the EU). Most cooperative and informal networks between different actors owe much to public subsidies and initiatives, though actors report that valuable cooperation may continue even if public funding has expired.

Public research funding encourages cooperation between different groups of actors, even if these actors are not too eager to cooperate. Therefore, some members of joint projects complained about others, e.g. saying that industry tries to usurp projects and asks for too frequent reporting where they are only expected to subsidize research, or that some universities profit from projects without offering all the results to industry, or that professors from the geotechnical sciences were largely not interested in publicly financed projects because they could earn more money via private orders for civil engineering projects. Cooperation within industry, i.e. between manufacturers and power supply companies, seems to be less contentious. Besides the international cooperation supported by EU projects, some major players cooperate internationally within their own corporate organisations. This is particularly useful if they have branches in the US and can benefit from the generous public research funding there.

The pre-combustion technology for carbon dioxide free power stations requires a special kind of cooperation because significant components consist of technologies which are not yet well known to power station operators, but are conventional in chemical engineering. Some actors anticipate challenges here with the necessary transfer of knowledge and cooperation: past experience has shown that chemical engineers and power station operators found it difficult to communicate and interact efficiently due to different cultures and different standards as, e.g. for standard piping diameters.

6.3.6 Bottlenecks and Opportunities

6.3.6.1 Opportunities

Most actors see carbon dioxide free power stations as a way of bridging the gap to the more distant future when hopefully renewable energy supplies and extensive energy efficiency efforts will be sufficient to secure energy supply while avoiding accelerated climate change. However, most actors do not expect carbon dioxide free power stations to be built during the next 15 years, a period during which a number of German power stations will need to be replaced.

In a broader perspective, the IGCC technology with CO_2 capture is also seen as a stepping stone to a hydrogen economy as this includes the production of hydrogen as part of the process. Instead of being subsequently combusted in the power station, some of it could be collected and transported elsewhere, e.g. to hydrogen filling stations. One interviewee even asserted that there will be no hydrogen economy without the gasification of coal, thereby opening the door to CO_2 capture.

Finally, carbon dioxide free power stations may become an export article if climate protection efforts are increased worldwide.

However, a number of obstacles and problems could be identified on the development path to carbon dioxide free power stations.

6.3.6.2 Technical Bottlenecks

Early Stage of Technology Cycle

The technologies for carbon dioxide free power stations are at an early stage of the technology cycle (phase 2) and according to most interviewed actors will not become serviceable on a large scale before 2020; possibly even much later, if ever. It is not yet clear which of the technology concepts will prevail and for what reasons. Generally speaking, all the technologies face the problem of not yet being efficient enough both in terms of energy and costs and are not yet ready for large-scale tests.

Some examples of technical bottlenecks are discussed in the following: for the pre-combustion route, there are no technologies available yet to burn hydrogen-rich gases in a gas turbine.[54] Furthermore, the overall low availability of the few existing IGCC power plants makes many actors doubt the reliability of the system set-up applied. To use post-combustion, the amine scrubbing technology needs to be improved considerably in order to reduce energy requirements and costs for the capture plant. The oxyfuel technology faces a number of problems, among them the need for a more efficient way of producing the necessary oxygen than the cryogenic method currently proposed. Oxygen transmitting membranes are one possible solution to this problem being investigated. On top of this, the combustion of fuels under the high temperatures of oxygen rich conditions or the alternative recirculation of the flue gas have not been resolved yet. Finally, the storage concepts are not yet fully developed either and may need long-running tests to gain the necessary level of certitude about their leak tightness. The existing geophysical methods are not accurate enough to provide evidence that at least 99.9% of the CO_2 is stored. This accuracy is necessary as leakage rates of even 0.1% of the stored mass will result in too large cumulated losses within a few hundred years. Compared with these fundamental problems, the issue of transport seems rather easy to tackle.

6.3.6.3 Other Bottlenecks

Costs

All interview partners agree that the single most important issue for the innovation system is the cost of producing energy with carbon dioxide free power stations. They expect additional costs of approximately 30 Euro per ton of carbon dioxide not emitted into the atmosphere to be reached in the medium-term future. Unless the price for emitting carbon dioxide reaches a certain level and is calculable in the long run, power supply companies will not build carbon dioxide free power stations. Of course, storage of carbon dioxide needs to be an accepted way to avoid buying certificates.

Additionally, reliability is an important point. Since it is very costly to build and run power stations, they must be a "proven technology" in the sense that it must be certain they will reach a sufficient level of availability.

Assuming that Vattenfall is successful with its pilot plant "Schwarze Pumpe", the construction of a demonstration plant would be the next logical step. It is still unclear which actor would be willing to make the necessary investments; most interview partners consider it indispensable to considerably increase public support in order to be able to build a demonstration plant.

Carbon dioxide free power stations will not only be expensive in terms of money, but will need substantially greater amounts of resources than conventional power stations. Since fossil fuel is expected to be in increasingly short supply in the coming decades, the volition to use some of these valuable resources to slow down climate

[54] More precisely, General Electric is the only technology manufacturer marketing gas turbines for operation with hydrogen-rich gases at present. However, as far as the authors are aware, there is not yet enough well-founded operational experience available.

change may decline. Therefore, some actors argue that power stations should be built in a way that makes it possible to turn carbon dioxide sequestration on and off. From today's perspective, this could be most easily achieved using post-combustion technologies.

Culture

The power supply industry used to be known as a "big tanker", changing direction very slowly, if at all. Many interview partners report a change here and now see Vattenfall, RWE and E.ON as being open to new developments. One actor, however, thought that RWE and E.ON in particular are still very conservative and very reluctant to change technology patterns.

Another aspect related to "technology culture" is the different approach to technical problems employed by mechanical engineers from the electricity industry and process engineers from the chemical industries. Both groups will need to collaborate more to realise power plants with CO_2 capture. Some interview partners confirmed the existence of these different traditions but it was not possible to determine the degree of friction that might arise in any necessary collaboration.

6.3.6.4 Public Opinion

All actors fear that public acceptance of carbon dioxide storage might be low, causing problems similar to the final storage of nuclear waste. Although this fear might be disproportionate, public acceptance of storage might turn out to be a bottleneck. Much depends on the public opinion on carbon dioxide free power stations in general, which still seems malleable at the moment.

6.3.6.5 Legal Definition of Carbon Dioxide

Finally, some interview partners pointed to the legal problems linked with carbon dioxide transport and storage. At the moment, the legal status of carbon dioxide captured from power plants is still unclear. If defined as waste, it is difficult or even impossible to transport it across borders and waste regulations will apply to its disposal. If defined as matter other than waste, it could be handled under the mining law when being stored. Many actors prefer the latter, while others ask for a separate, specific legislation.

6.3.7 Policy Recommendations for Public Research Policy

Certainly not all of the problems connected with innovation in the field of CO_2 free power plants can be solved by political measures and only some can be addressed by public research policy. Nevertheless, some specific conclusions and recommendations can be formulated for the innovation system and more generalised conclusions also drawn.

First of all, the innovation system for CO_2 free power stations is very active and could certainly be described as being in the phase of euphoria. As mentioned above, the term "euphoria" refers to the concept of the technology cycle and should not be misunderstood as implying that the actors themselves are euphoric. Nevertheless, the amount of activity reported in the interviews has been constantly increasing while maintaining a diversity of approaches and there are industrial players inside and outside Germany accelerating the pace of development by announcing projects approaching industrial dimensions. From a public perspective, the development is well underway as most of the industrial players of the innovation system have adopted the technology concept of CO_2 capture and storage in their strategies. If these strategies are upheld, they would actually enable the electricity sector to significantly reduce emissions and thereby contribute to meeting the political goal of emission reductions. Concerning the reported amount of funds and level of activity, a further fundamental improvement would only be possible by increasing the public funds by several orders of magnitude.

6.3.7.1 Public Research Funding

Given the early stage of the technology cycle, different technology routes should be pursued at the moment. However, future results may provide the information necessary to focus research funding. Therefore, monitoring on a regular basis might be useful to be able to react quickly to any new research results and use them to focus public funding as soon as possible, thereby speeding up the process. Such monitoring could take place in the way the COORETEC process started, i.e. by bringing together all the stakeholders under the chairmanship of the BMWi and then allowing these players themselves to identify the promising technologies and their relevant bottlenecks. The whole value chain should be taken into account in this process, including capture, transport and storage.

Although most actors are more or less satisfied with the way public research funding is allocated, it could be improved by bringing together the responsibility for all aspects of CO_2 capture and storage in one ministry, thus lowering transaction costs. Such a centralisation would make industrial actors confident that all steps in the value chain of CO_2 capture and storage are being given a comparable amount of attention and that the activities are target-oriented. The latter is especially important for the storage side, where solid knowledge about realisable storage potentials is of crucial importance and not the investigation of entirely new hypothetical storage concepts. Given the low chances of success in being able to restructure the responsibility of federal ministries, at the very least, the cooperation between ministries should be improved by regular and short-term mutual exchanges of the competent officials of plans, programmes and individual projects.

The fact that a first pilot plant and an industrial scale plant have been announced in Germany and in the United Kingdom, respectively, underlines the industries' need for rapid progress in this field. As there are still high technical risks in these plants, these initiatives can be characterised as leapfrogging projects. The policy system and also the academic research system in Germany, which is financially strongly dependent on policy induced funding, are not equipped for such leapfrogging action. This is true despite the fact that leapfrogging meets the political demand of developing available solutions quickly in order to tackle climate change. Hence, the research programmes on carbon capture and storage should actively foster such leapfrogging instead of being restricted to the ideal model of moving from fundamental basic research through applied research to market diffusion.

6.3.7.2 Focus Public Research Funding in Time

The present research activities in general and also the publicly funded activities are still of a rather wide scope, a fact that is in line with the status of CO_2 capture and storage as being in a very early phase of the innovation system. As understanding and knowledge increase as a consequence of the research activities, the attempt should be made to identify and select promising solutions worth more concentrated research efforts. This identification and selection process could be performed by a joint stakeholder group such as the COORETEC-Council. However, adequate care should be taken to balance out the interests of individual actors in the innovation system.

6.3.7.3 Formulation of Strategies and Roadmaps

Strategies and road mapping studies would have to go beyond the exercise performed within the initial phase of the COREETEC programme, where a roadmap for future fossil fuelled power plants was developed. This roadmap largely reflects the mid-term strategies of the technology and electricity producers involved, but does not give an integral view of the required emissions reduction of the entire energy system and of the contribution of the electricity system to meeting this requirement.

The activities of the Swiss Energy Research Commission (CORE) are one example where an integral road mapping exercise is being undertaken (compare Bürer and Cremer, 2006). In particular, the 2000 Watt per capita society establishes a clear vision of the future. Based on this vision, concrete technology requirements can then be derived, needed to satisfy a future energy service demand while meeting the requirements of the 2000 Watt per capita society. The technology requirements in turn lay the foundation for formulating research needs and also improve the communication of requirements for long-term policy development.

6.3.7.4 Foster National Cooperation

Fostering cooperation should be an integral part of any research policy or any activity of research programme development. In the context of CO_2 free power stations, all actors underlined the effectiveness of the work performed when outlining the research programme COORETEC and also the fruitfulness of research cooperation under the framework of the "AG-Turbo".

6.3.7.5 Foster International Cooperation and Evaluation

Astonishingly, some parts of the German innovation system are still more or less entirely oriented towards a national framework. This means that there are actors in the research system who do not cooperate voluntarily with international partners and sometimes have not done so at all. Furthermore, the German research projects and programmes are not subject to an "external" evaluation of experts outside the German innovation system. In view of the fact that CO_2 free power stations and most other energy related research topics are being discussed all over the world and that the technology industries as well as the electricity industry to a growing extent are internationally oriented, international cooperation makes sense and could help to prevent the duplication of work. Besides international cooperation, external evaluation by foreign experts would help to improve the quality of research activities and perhaps facilitate the adaptation and adoption of strategies that have already proven successful abroad.

Some interviewed actors showed opposition to the need for international cooperation in EU-projects based on the argument that these often involve partners of limited expertise who have to be integrated in order to be eligible for project funding. Although this may occur, the research policy actors should still try to enhance international cooperation because German actors will only be able to build networks of highly qualified members via international cooperation. Such networks in turn could act on the international "market" for research activities and transfer knowledge back to the domestic innovation system.

6.3.8 Other Policy Recommendations

6.3.8.1 Role of Research Policy in the Context of European Energy Policy

The single most significant bottleneck to building carbon dioxide free power stations is the uncertainty about the future costs for CO_2 capture and storage and the future costs of carbon dioxide emissions. Therefore, the most important political measures necessary are clear decisions about the restriction of carbon dioxide emissions on the global, European and national levels. Such decisions certainly go far beyond the scope of research policy and it would not make sense to demand them of the policy actors in the innovation system. However, an improvement could be reached if actors from research policy were to see themselves as a pivotal hub between the technological innovation system and the overall policy system and acted accordingly. Of course, the individual actors from the administrative levels of ministries and supporting agencies do not influence higher policy to the same degree as lobbying actors from industry. However, as these authorities are responsible for developing strategic views and documents on the future energy system, they do have access to an instrument for identifying and efficiently communicating the framework needs of a technological innovation system such as the one for CO_2 free power stations. Moreover, such publicly developed strategic documents have a higher credibility than the views communicated by industrial actors of the innovation system.

6.3.8.2 Legal und Public Aspects

Legal definitions are necessary to deal with the concerns of actors about the storage of carbon dioxide. Such definitions make it possible to adequately regulate transport and storage and address the interests of both industry and environment groups. All stakeholders need to be well informed and should be included in the legislative process in order to prevent massive public concern as is the case with nuclear waste. An information campaign – whether funded by public or private institutions – which shows the use of CCS as one way to reduce greenhouse gas emissions might also help to overcome the concerns of utilities about possible public resistance to carbon dioxide storage.

6.3.8.3 Ensure Enough Manpower

Since there is no consensus among interview partners about the future availability of qualified personnel, this issue should be analysed with care. Although it is hardly possible to anticipate how many engineers with what qualifications will be needed to build and run carbon dioxide free power stations, it should be ensured that the research programmes contribute to the relevant education/training programmes at German universities.

References

Arbeitsgemeinschaft Energiebilanzen (2006). www.ag-energiebilanzen.de

Buller, A. T., Kårstad, O., & de Koeijer, G. (2004). Carbon Dioxide – Capture, Storage and Utilization, Statoil Research and Technology Memoir No. 5, Stavanger, obtainable through contact: ktjmn@statoil.com

Bundesministerium für Wirtschaft und Arbeit. (Hrsg.). (2003). Forschungs- und Entwicklungskonzept für emissionsarme fossil befeuerte Kraftwerke. Bericht der COORETEC Arbeitsgruppen. Dokumentation Nr. 527 des BMWA

Bürer, M., & Cremer, C. (2006). *A contribution to the identification of promising technologies for the 2050 Swiss energy R&D policy vision*. Schlussbericht. Bundesamt für Energie, Bern

Calabro, A., Deiana, P., Di Mario, F., Fiorini, P., Girardi, G., Iacobazzi, A., et al. (2005, January)."HYPOGEN Pre-feasibility study", Report EUR 21512 EN.

Christensen, D. (2004). CO_2 Capture Project's Policies and Incentives Study, Presentation at the CO_2 Capture Projects Phase 1 Results Workshop, Brussels, June 2nd 2003.

Göttlicher, G. (1999). Energetik der Kohlendioxidrückhaltung in Kraftwerken. VDI Fortschrittsberichte, Reihe 6 Nr. 421. VDI Verlag, Düsseldorf.

International Energy Agency IEA. (2005).*CO_2 Emissions From Fuel Combustion 1971–2003*. Paris.

IPCC. (2005). IPCC Special Report on Carbon Dioxide Capture and Storage, Prepared by Working Group III of the Intergovernmental Panel on Climate Change. Metz, B., Davidson, O., de Coninck, H.C., Loos, M., Meyer, L.A. (Eds.). Cambridge, New York: Cambridge University Press.

Radgen, P., Cremer, C., Warkentin, S., Gerling, P., May, F., & Knopf, S. (2005). Bewertung von Verfahren zur CO_2-Abscheidung und -Deponierung. Fraunhofer Institut für Systemtechnik und Innovationsforschung (ISI), Karlsruhe und Bundesanstalt für Geowissenschaften und Rohstoffe, Hannover, Abschlussbericht an das Umweltbundesamt (in Veröffentlichung).

Radgen, P., Cremer, C., Warkentin, S., Gerling, R., May, F., & Knopf, S. (2006). Bewertung von Verfahren zur CO_2-Abscheidung und -Deponierung. Research Report 203 41 110. Umweltbundesamt, Dessau.

Shackley, S., McLachlan, C., & Gough,C. (2005). The public perception of carbon dioxide capture and storage in the UK: results from focus groups and a survey. *Climate Policy, 4, 377–398.*

VDEW. (2004). *VDEW-Jahresbericht 2004*. Berlin: Verband der Elektrizitätswirtschaft.

VDEW. (2006). Stromwirtschaft nutzt vielfältigen Energiemix (aus: www.strom.de).

Markewitz, P., Martinsen, D., & Vögele, S. (2004). The Future Role of CO_2-Capture as Part of a German Mitigation Strategy. STE-Reprint 19/2004. Programmgruppe Systemforschung und Technologische Entwicklung (STE). Forschungszentrum Jülich.

Chapter 7
Industrial Furnaces

7.1 Technology Description and Energy-Economic Significance

Industrial kilns and furnaces consume about 30% of the fuels used by industry, mostly in basic product industries, and also some 10% of the electricity demand (e.g. for secondary steel, primary aluminium, glass (for additional heat) and high quality ceramics). However, the types of kilns and furnaces differ substantially in detail regarding the heat source (burners, electric resistance and electric arc), the design of the containment, the type of operation (batch or continuous), the insulation and the control techniques. Table 7.1 gives an overview of this variety, listing typical products together with the types of furnaces often used. Due to this variety – also expressed in the different levels of temperatures, different sizes and different demands on the specification of the insulation material – and due to the fact that most of the applications of kilns and furnaces are traditional technologies, there are many opportunities for (minor, incremental) improvements, but few for major new technologies and technical breakthroughs.

This variety has several consequences for the analytical work in this research project:

- most kilns and furnaces are positioned in the diffusion phase of the technology cycle, and only certain elements of the kiln or furnace system are the subject of research and patenting. The systems as a whole often represent mature technologies; only individual components are undergoing significant improvement or fundamental breakthroughs;
- as a result, the question arises as to where and to what extent public R&D should be involved in promoting technological progress. Depending on the type of furnace, the number of applications in the EU can be rather limited or quite widespread. Therefore the question arises whether R&D for this technology should be targeted by national or EU R&D programmes and demonstration projects.

Due to the considerable variety of kilns and furnaces, it was concluded that this analysis should concentrate on a limited number of furnaces, and the decision was

E. Jochem (ed.), *Improving the Efficiency of R&D and the Market Diffusion of Energy Technologies*, DOI 10.1007/978-3-7908-2154-3_7, © Physica-Verlag Heidelberg 2009

Table 7.1 Industrial furnace types and their field of application, fuel and electricity demand in Germany 1992

		Energy demand in PJ (1992, Germany)		
Product	Type of furnace most often used	Fuels	Electricity	Total
Pig iron	Blast furnace	284.7[1]	31.63	316.3
Sinter	Sintering furnace	38.1	4.23	42.3
Coke production	Shaft furnace	67.0[3]	3.0	70.0
Oxygen steel	Melting furnace	20.2	2.24	22.4
Electric steel	Crucible furnace	6.4	14.92[2]	21.3
Rolled steel	(Re-)heating furnace	60.5	6.72	67.2
Iron and steel foundries	Cupola furnace, crucible furnace	16.9	4.78[2]	21.7
Remelt Al	Melting furnace	2.4	0.61[2]	3.1
Copper, zinc, lead	Rotary furnace, pusher furnace	7.8	0.35[2]	8.2
Non-ferrous semi-finished products	(Re-)heating furnace	14.5	9.63[2]	24.1
Non-ferrous metal foundries	Melting furnace	5.7	2.2[2]	7.9
Olefins	Tube furnace	86.2[4]	0.87	87.1
Alumina	Calcining furnace	2.5	0.35	2.9
Cement	Rotary kiln	74.9	6.51	81.4
Bricks and tiles	Tunnel kiln	10.7	0.56	11.2
Lime	Shaft furnace	24.7	1.3	26.0
Other non-metallic minerals	Chamber oven	19.6	1.71[2]	21.3
Industrial goods	Heat treating furnace	26.9	14.71[2]	41.6
Glass	Tank furnace	54.1	6.02[2]	60.2
Ceramics	Tunnel kiln and chamber oven	9.7	0.51[2]	10.2
Bakery products	Tunnel oven and chamber oven	8.2	0.91[2]	9.1
Sum total		840	115	955

[1] Part of fuel used as reducing agent.
[2] The electricity demand given here is used mainly to generate thermal energy.
[3] Consumption in transformation sector of German annual energy statistics (Energiebilanz of the FRG).
[4] Listed under non-energetic consumption in the Energiebilanz. of the Fed. Republic of Germany.
Source: Radgen, 1998.

made to examine the various furnaces used in the iron and steel industry. In addition, we focused on furnaces which can be used as a cross-cutting technology rather than on specialized furnaces which are individually designed based on the reactions taking place in the furnace such as the blast furnace.

7.2 The Analysis on a Micro-Level

Although industrial furnaces and kilns are traditional technologies, improvements in reducing emissions and efficient energy use remain a top priority beside improving the properties of the products. These technologies have matured in industrialised countries during the last century, but their major investment is shifting to emerging countries like China, India and countries in South-East Asia and Latin America. These characteristics seem to be reflected in the indicators of the following two sections.

7.2.1 Technological and Economic Challenges

Looking back over the last 15 years, two main technological challenges can be identified in the area of industrial furnaces and components. One is related to nitrogen oxide emissions from combustion and the second to the carcinogenic impacts of fibrous insulation materials used in high temperature furnaces.

The German technical directive on air quality control (TA Luft) sets emission limits for nitrogen oxide emissions from fossil fuel combustion plants. Around 1990, these limits were significantly reduced, making it necessary to think about end-of-pipe technologies for exhaust gas cleaning. The catalytic conversion of nitrogen oxide was considered but this would have caused significant additional costs for equipment and consumables. This pressure then became the driving force for developments to reduce the NO_x emissions at source, mainly the burner. New burner types (e.g. FLOX burner) and new combustion strategies with recycled flue gas were developed. The aim of these developments was to reduce the combustion temperature and to optimise the combustion so that the NO_x emissions would be reduced to such an extent that the emission limits could be met without the need for expensive, additional end-of-pipe technologies.

Ideally, high temperature furnaces should be well insulated and have a low heating capacity to enable a fast modification in furnace temperature. Fibrous insulation materials (materials besides asbestos) had been widely used in furnaces but were identified as having carcinogenic impacts in the 1990s. The use of such fibrous insulation materials had a positive influence on the energy consumption of the furnace, as the furnace temperature was able to be altered rapidly. This made it possible sometimes to turn off the furnace or reduce its temperature over weekends or even overnight. However these original insulation materials had to be replaced by others (typically with longer fibres) which were developed to replace those no longer acceptable for health reasons.

The main economic challenge in furnaces is the high energy cost. Most furnaces are operated using natural gas or electricity, and generate high energy costs during operation. Furnaces are typical for the product group in which operating costs (which are mainly energy) dominate life cycle costs. Therefore the continuous improvement of energy efficiency has always been a major issue in the sector.

The market for furnaces and burners has grown over the last nine years by more than 50%, but with significant variation in market size by year. The reason for this fluctuation might be the importance of the process plant construction, especially in the iron and steel industry.

The world market for industrial furnaces is highly concentrated; the five market leaders (Germany, Italy, United States, Japan, and France/United Kingdom) held a market share of about 70% between 1995 and 2000, see Fig. 7.1. During this period, France and the UK changed positions in the system. This was mostly influenced by the exchange rate of the British Pound in relation to the Euro, making UK products more or less economic on the market.

Since 2000, the total market share of the five leading countries has been decreasing and now amounts to 62.4% of the world market (see Fig. 7.2). Some parts of the world market were lost, not to other European competitors, but to three main countries: Russia, Korea and China. These had been able to increase their market share from 0.8% in 1995 to 6.1% in 2004, an increase of more than 600%. It is obvious that these new market entrants may become strong competitors on the world market, resulting in a further drop in the market share of the five previous leading countries. This will probably lead to a further decline in prices and greater competition, forcing companies to merge. Therefore a concentration process in each country may occur at the same time as new competitors on the market appear. However these changes seem to occur slowly, which makes adaptation by industry smoother.

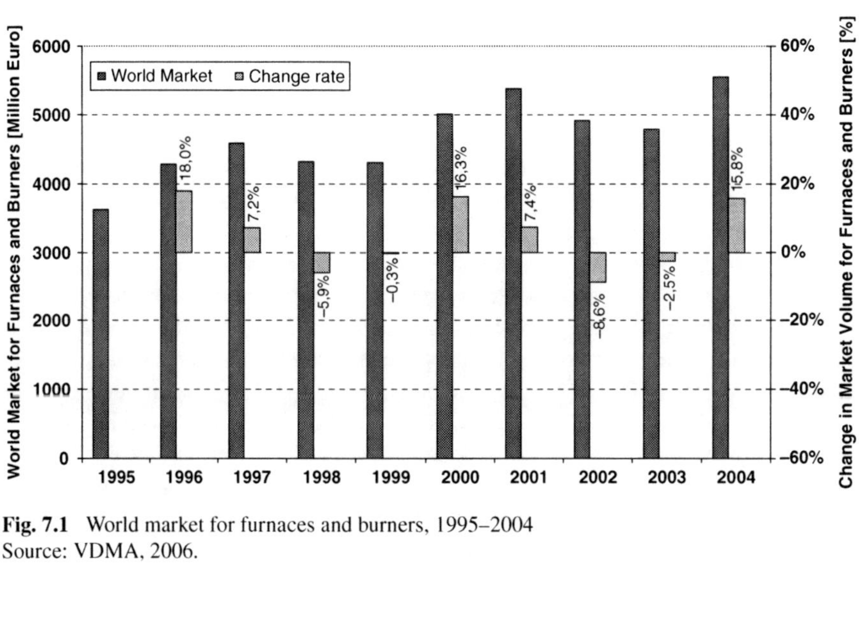

Fig. 7.1 World market for furnaces and burners, 1995–2004
Source: VDMA, 2006.

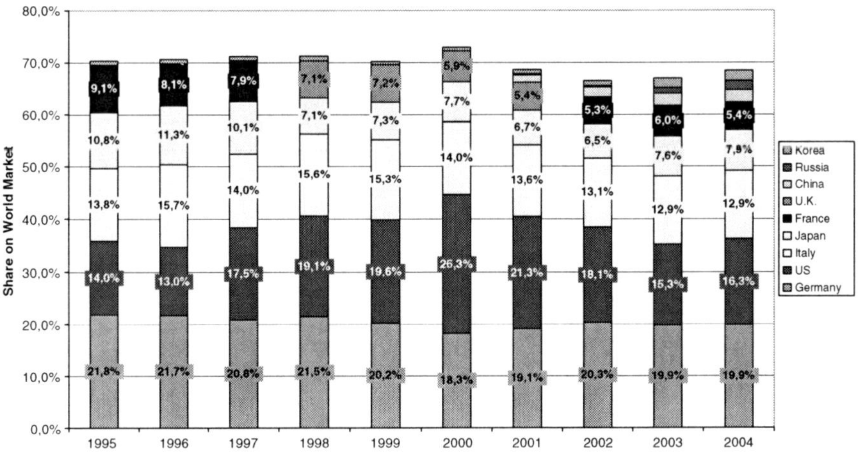

Fig. 7.2 Market shares of market leaders and important new competitors in the field of furnaces and burners (Data source: VDMA, own calculations), 1995–2004

For the German furnace industry, which is dominated by SMEs, one of the main economic challenges is to remain competitive on the world market where customers are typically large multi-national companies.

The challenge can best be visualised by looking at the Revealed Comparative Advantage (RCA) indicator, which declined between 1995 and 1999 (see Fig. 7.3).

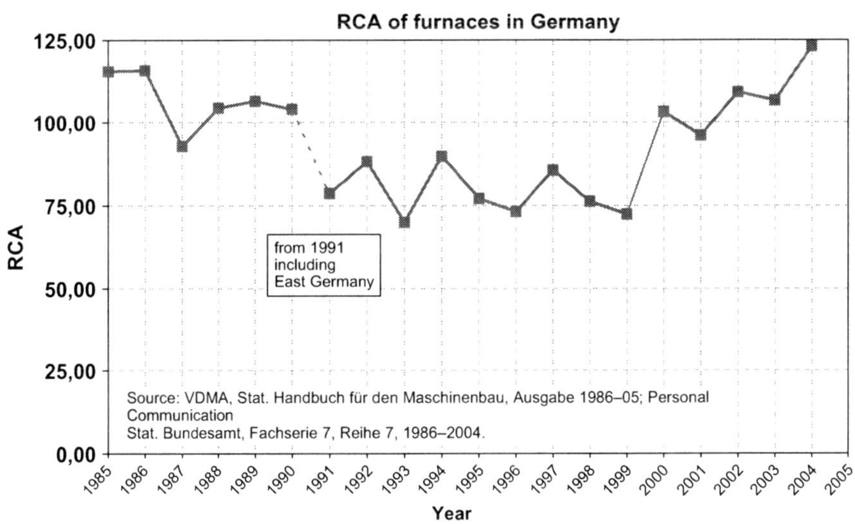

Fig. 7.3 Revealed Comparative Advantage for furnaces in Germany, 1985–2004

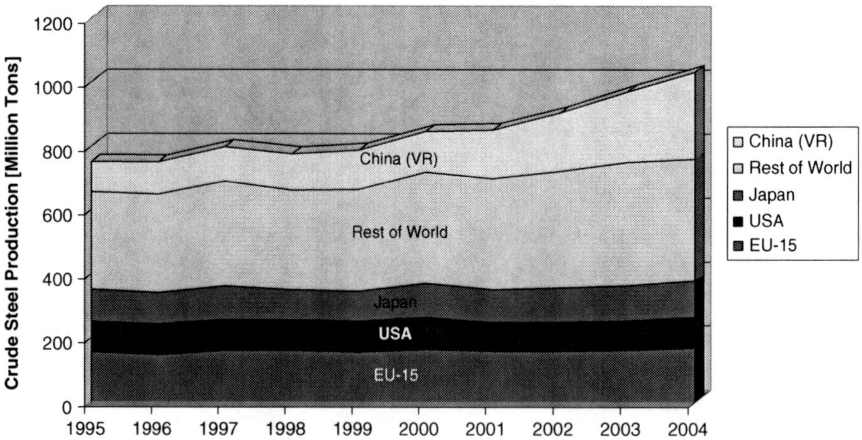

Fig. 7.4 Crude steel production between 1995 and 2004
Source: Statistische Jahrbücher der Stahlindustrie, Hrsg.: WV Stahl, Stahlinstitut VDEh.

Since 2000, a strong increase of the RCA value can be identified, which requires interpretation. The reason behind this significant switch from decreasing values to increasing values is related to developments in the steel industry. Since 1999/2000, the production of crude steel has been increasing strongly and continuously. However, this increase has mainly taken place in China (see Fig. 7.4).

An RCA value above zero describes a better than average competitive position relative to the other sectors of German industry. The trend shows declining competitiveness from more than 100 in the second half of the 1980s to 65 in 1999–2001 which raises questions about the possible explanations of this development.

7.2.2 Position in the Technology Cycle

7.2.2.1 Patent Dynamics

To identify the phase of the technology cycle, it is crucial to analyse the dynamics of patent applications. The industrial furnace patents refer to the furnace as a complete system and the components that comprise it. The total number of patent applications related to furnaces or components used in furnaces has been slowly increasing over the last 20 years (see Fig. 7.5).

A first look at the patent applications by area from 1985 to 2003 reveals that there has been a substantial increase in patenting burners due to the additional environmental obligations mentioned above. After the rapid oil price decline in 1985, the 10 year decline in patents for insulation materials seems to be understandable before their increase again over the past few years due to substantial fuel price increases. The numbers for combustion furnaces also increased, possibly due to changing waste disposal legislation in most European countries (no waste disposal

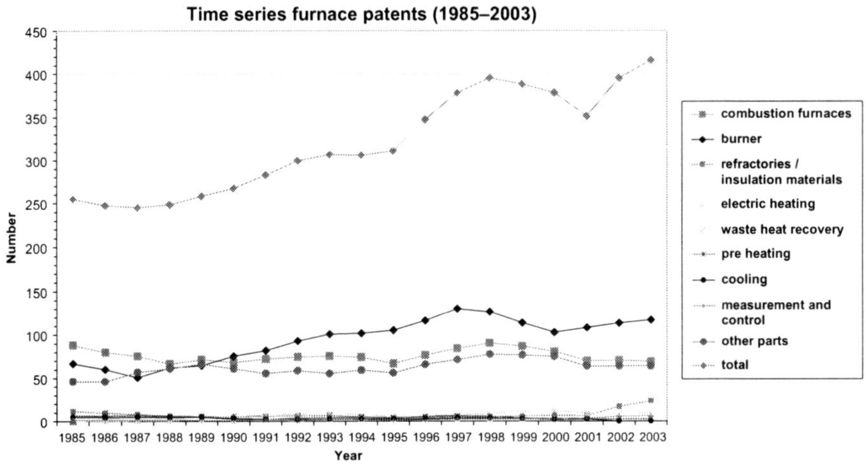

Fig. 7.5 Number of patent applications related to furnaces at the EPO, Germany, 1985–2003

of burnable wastes) and the growing interest in burning wood chips and wood wastes in large boilers.

In general, industrial furnaces are mostly at the end of the diffusion phase of the technology cycle, where only a few completely new ideas are developed, but additional applications may still evolve (e.g. wood burning furnaces). As the differences in numbers are not very large, the following two figures distinguish two blocks of patent applications: the first comprises those with more than 50 patent applications per year (Fig. 7.6) and the second those with less than 50 patent applications per year (Fig. 7.7).

On closer examination, it is possible to identify certain dynamics: burners were the most dynamic technological area measured by patents for industrial furnaces from 1990 on. This is not surprising when it is taken into account that the FLOX-burner and the recuperative burner were developed at this time. These burner developments were triggered by more stringent NO_x emission limits which could finally be met with new or improved burner designs. The two other groups, of which one is the furnace as a whole, remained very stable over a very long time, which is a good indicator of a technology at the end of the technology cycle with some expectation of new applications (such as wastes and wood).

In patents with lower numbers per year, the visible trends can be explained by several external effects which are driving the furnace technology or its elements (see Fig. 7.7). However, care must be taken when interpreting these figures given the very small numbers of patents involved; changing slopes or directions might easily be overestimated and interpreted as a trend when they may actually be the result of simple variability.

An area in which there has been a rapid increase in patent applications, especially since 2002, is insulation materials/refractories. Although the overall number of patents is still rather small, this is interesting because it matches statements made

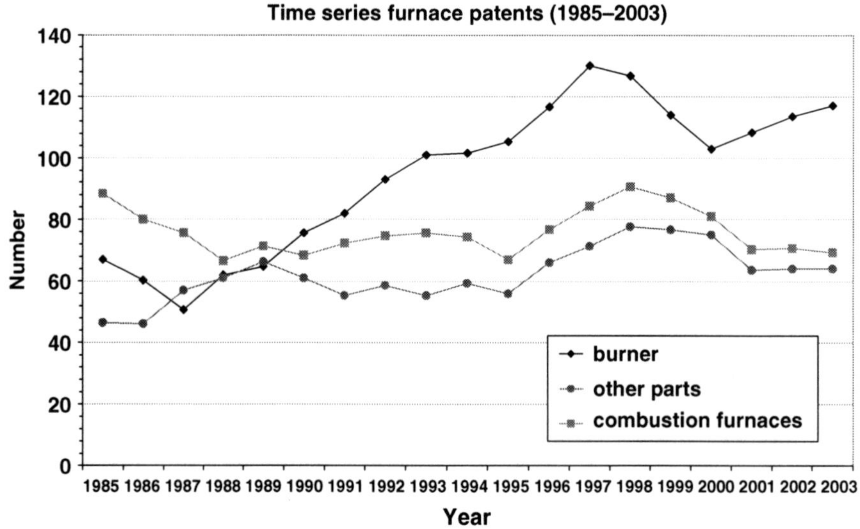

Fig.7.6 Number of patent applications in relation to furnaces (areas with higher activities of patenting), Germany, 1985–2003

by our interviewees about possible promising fields of R&D in the near future. Not as explicit, but still worth mentioning, is the development in patents for waste heat utilisation which indicate a way of increasing energy efficiency which has become more and more popular over the last few years. Both trends are probably based on

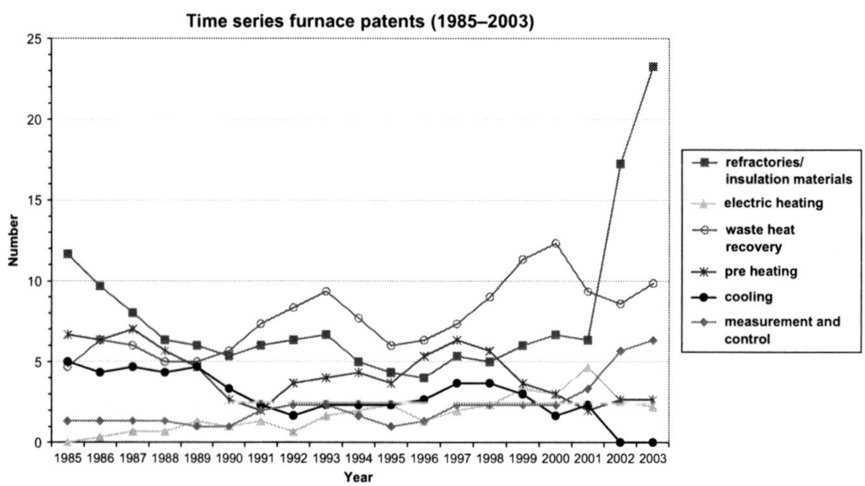

Fig. 7.7 Number of patent applications in relation to furnaces (areas with lower activity), 1985–2003

the energy price increase since 2000 which is likely to continue to be supported by rather high fuel price levels in the near term.

Overall, three trends can be identified:

1. Decreasing number of patents on cooling
(→ better insulation materials make cooling redundant).
2. Increasing numbers for measurement and control
(→ achieving high and reliable product quality is the major challenge faced by furnace operators. Exact measurement and control is essential to guarantee the desired parameters in the furnace).
3. Increasing importance of insulation material.

Figure 7.8 supports the previous statements. There are few areas of dynamic development apart from the growing importance of burner-related patents and the rapid growth in refractories' patents. However, looking at the share of specific patent classifications compared to all relevant patents reveals that waste heat utilisation is stronger in 2003 than it was in 2002 and nearly as strong as in 2000, but that this has fluctuated constantly over the years. Another interesting finding is the fact that burner-related patents dominate the patent statistics, which highlights the significance of the burner for the furnace as a whole. However, burner development is often pushed by the manufacturers of power generation equipment. A gas turbine combustor has many things in common with a burner for an industrial furnace and NO_x emissions have also been a major concern for gas turbines.

Figure 7.9 confirms the "stability" in a geographical sense. It shows that the distribution of originating countries remained stable between 1990 and 2001. Furthermore, it indicates the strong position of Germany, although the gap to Japan has narrowed somewhat. Astonishingly, France ranks fourth on the list, even though France lost its position in the top 5 world exporters for furnaces in 1998 and was replaced by Great Britain, which ranks far below it in the number of patent applications.

To obtain more detailed insights into which actors are driving new technologies or the improvement of existing ones, the lists of the top 10 applicants were obtained for different years. Table 7.2 summarizes the data for the years 1988, 1990, 1994, 2000 and 2001. The companies on this list can be mainly linked to three different activities:

- Steel Industry (Mannesmann, Nippon steel, Kawasaki steel, Nisshin steel, JFE steel, Mitsubishi heavy industry, Voest-Alpine)
- Cement/Glass Industry (Krupp Polysius, Sorg, Schott, Carl Zeiss)
- Power Generation (ABB/Asean Brown Boveri/ABB research, Babcock –Hitachi, General Electric, Alstom Technology, Siemens AG)

Whereas the first two groups represent user industries in which the overall process incorporates many furnace applications, the third group represents the manufacturers of power generating equipment. In sectors where improvements and therefore

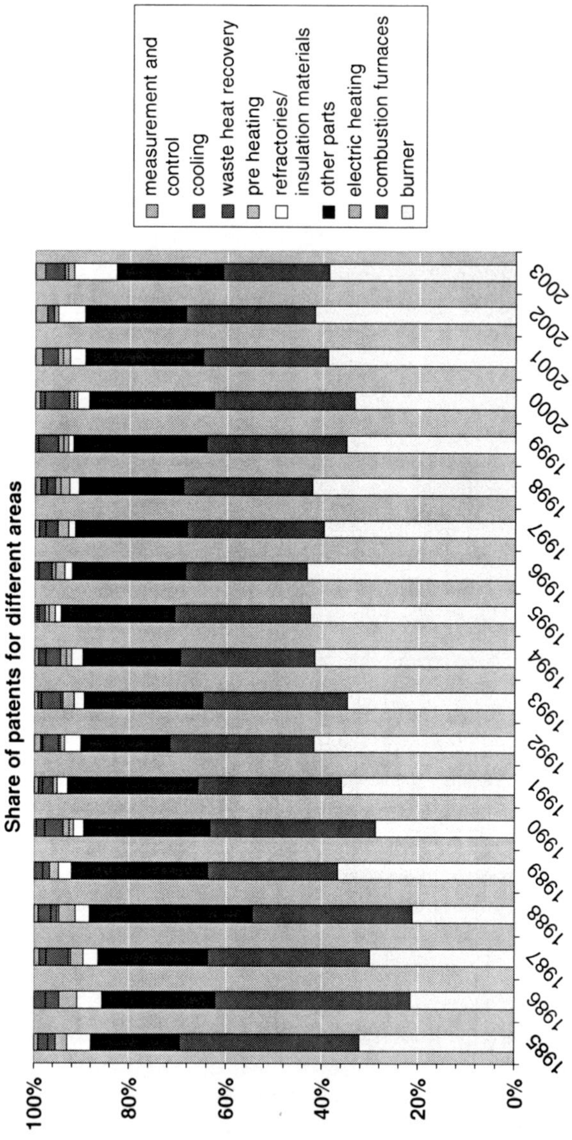

Fig. 7.8 Share of patents for different countries, 1990 and 2001

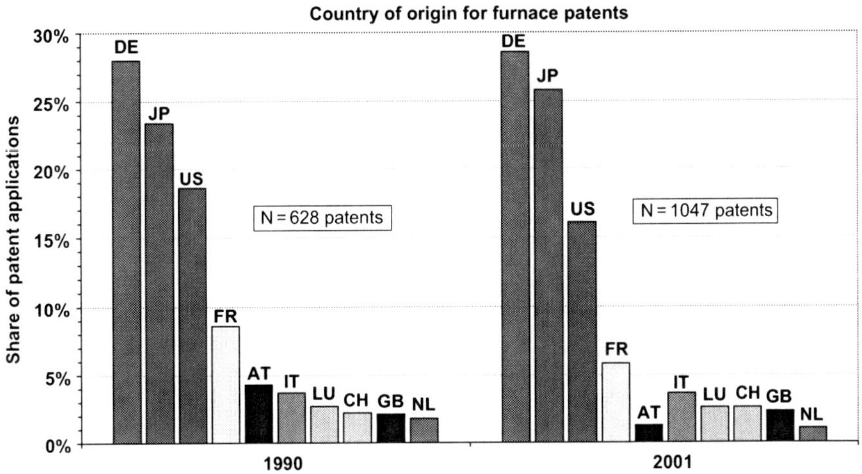

Fig. 7.9 Country of origin for furnace patents, 1990 and 2001

the basis for patents are developed mainly in the industries in which the technology is applied, it can be stated that they have already arrived at the end of the technology cycle.

This finding was also confirmed by obtaining the numbers of patent applications of the identified key actors in the field of industrial furnaces, Table 7.3. Again R&D

Table 7.2 Top 10 patent applicants, 1988–2001

	1988		1990		1994		2000		2001	
No	**Applicant**	Number of patents	**Applicant**	Number of patents	**Applicant**	Number of patents	**Applicant**	Number of patents	**Applicant**	Number of patents
1	Mannesmann Aktiengesellschaft	6	Krupp Polysius AG	7	ABB Managment AG	10	JFE Steel Corporation	25	Alstom Technology Ltd.	11
2	Union Carbide Corporation	6	Kawasaki Steel Corporation	6	Asea Brown Boveri AG	10	Schott Glas	16	Alstom (Switzerland) Ltd.	10
3	NGK Insulators Ltd.	5	Asea Brown Boveri AG	5	Nippon Steel Corporation	10	Kawasaki Steel Corporation	15	JFE Steel Corporation	10
4	Nippon Steel Corporation	4	Paul Wurth SA	5	ABB Research Ltd.	8	Nippon Steel Corporation	11	SMS Demag AG	10
5	Praxair Technology, Inc.	4	VOEST-Alpine Industrieanlagenbaugesellschaft GmbH	5	Krupp Polysius AG	7	Carl Zeiss Stiftung trading as Schott Glas	10	Nippon Steel Corporation	9
6	Stein Heurtey	4	BGK Finishing Systems, Inc.	4	Air Liquide SA	6	Alstom Technology Ltd.	9	Paul Wurth SA	9
7	Veitscher Magnesitwerke-Actien-Gersellschaft	4	Didier-Werke AG	4	Nisshin Steel Co., Ltd.	6	Paul Wurth SA	9	Schott AG	8
8	Allegheny Ludlum Corporation	3	Applied Materials, Inc.	3	Praxair Technology, Inc.	6	Alstom (Switzerland) Ltd.	8	Mitsubishi Heacy Industeries Ltd.	6
9	Beteiligungen Sorg GmbH & Co KG	3	Babcock-Hitachi K.K.	3	General Electric Company	5	General Electric Company	8	Siemens AG	6
10	Didier Engineering GmbH	3	Eaton Corporation	3	VOEST-Alpine Industrieanlagenbau GmbH	5	Outokumpu Oyj	8	Webasto Thermosysteme International GmbH	5
	Total Top 10	**42**		**45**		**73**		**119**		**84**
	Total all	309		336		410		479		412
	Share of Top 10	13,6%		13,4%		17,8%		24,8%		20,4%

Table 7.3 Number of patents of some key actors, 1988–2001

	Number of patents						
	1988	1990	1994	1998	2000	2001	Total
Industry							
Maerz Gautschi					1		1
SMS Demag			4	9	7	10	30
LOI Thermprocess		3	2		2	2	9
Riedhammer			3				3
Aichelin		2	1	1			4
Ipsen	1					3	4
Nabertherm							0
Ebner			1				1
Voest-Alpine	3	5	5	2	2	2	19
Krupp Polysius	3	7	7		2		19
IOB Industrieofenbau							0
Ruhstrat							0
Techint Italimpianti							0
Otto Junker	1		1	2		2	6
SKS							0
WS Wärmeprozesstechnik	1						1
DBK David + Baader							0
Kromschröder							0
Crone							0
LBE		1					1
F&S Feuerfestbau							0
Research							
GWI Gaswärme Institut							0
VdEh-BFI							0
RWTH							0
FHG			1				1
Uni Bochum, LEAT							0
TU Freiberg							0
Dt. Inst. f. Feuerfest und Keramik, Bonn							0
MPI							0
TU Illmenau							0
Forschungsgemeinschaft Feuerfest							0
Uni Bremen							0
TU Clausthal					1		1
TU Karlsruhe							0
Engler-Bunte-Institut							0
FZ KA							0
Uni Magdeburg							0
Total	9	18	25	14	15	17	98

Source: European Patent Office.

institutions show up with as few as two patent applications, whereas industrial companies applied for 96 patents during the years analysed.

To evaluate these results concerning the key actors, we decided to try and verify them by a kind of backward-analysis. To do so, one of the industrial key actors provided an overall compilation of their patents. According to this, they have a number of accorded or pending patent applications at the European Patent Office to date. Based on our results in Table 7.3, our research covered 60% of them. The question thus arises why our analysis did not manage to include the other 40% of their patents.

To answer this question, we undertook a second analysis searching explicitly for the patents of this special key actor and compared these results with the prior ones. First, all of the patents mentioned in the company's compilation were retrieved, so it was possible to then examine why some of them did not appear in our initial analysis.

Twenty-seven percent of the patent applications not covered are not within the period of time under observation, so it is obvious why they were not accounted for. Similarly, 7% are in a patent classification that was deliberately excluded from the analysis. The remaining 7% of patents actually should have been covered by our initial research, because they match both the period analysed as well as the selected patent classes. So there is no reasonable explanation why these particular patents were not captured by the original research strategy. Maybe this phenomenon is the result of a database error or inconsistency.

Concerning the overall performance of the applicant analysis conducted here, it is possible to reach the following conclusions: the analysis proved viable with reference to the technological perspective. With one exception, all patents were included which matched the selected classifications as well as the selected years. However, should the main focus be placed on key actors' activities, it might be necessary to rethink the analytical strategy. As our test has shown, it is useful to then search systematically for the applicant's name and not primarily for pre-selected years in order to obtain a complete picture of the actor's activity. Secondly, and this is a general problem of patent analyses, there is the question of whether the search should be narrowed down by focussing on specific patent classifications or not. If this is done, important outcomes might be systematically excluded. If, on the other hand, the search is conducted more broadly, there is the risk of broadening the analysis and obtaining diffuse results, or even worse, mis-measurements. Closely related to this point, selecting which database(s) to use is another crucial decision the researcher has to make. Depending on the individual actors and their main markets, it might be necessary to look at national patent offices, too. Possible name changes of actors over the course of time also have to be taken into account. The search strategy must also consider company merges and other similar developments and be adjusted accordingly. For these reasons, it is essential to obtain reliable information about the actors of interest.

A final aspect concerns a structural problem of data. Due to patent application procedure, there is a time lag between the initial application and the publication of this application (for example 18 months at the European Patent Office). As a result, current patents that have been applied for in the last two years cannot be analysed because they are not yet listed in the databases.

A further indicator for innovation dynamics in a technology field is the analysis of patent citations. The analysis of patent citation has two dimensions, one looks forward and one backwards. The forward-looking analysis examines the number of citations of relevant patents in subsequent patents. The analysis of forward citations shows the relevance of the patent of a specific year in the following year. Important innovations should show up as an increase in patent citations in the following years.

Therefore a technology at the end of its technology cycle should show decreasing numbers of citations in the short (0–4 years) and in the long term (0–14 years) since the most important basic patents were already claimed many years before. However the problem is that the database only covers the years from 1985 to 2001, therefore the citation numbers are partly incomplete. The 90 patents on combustion furnaces in 1985 were cited 24 times in the next four years (1985–1989) and 63 times in the next fourteen years (1985–1999), cf. Table 7.4. The average citation rate stays around a constant value, indicating that key patents were made before the analysed period, which is a typical indicator for a mature technology.

An example of the backward citation index for burners is shown in Table 7.5. This is able to exploit a much more complete set of data due to the better coverage of the time horizon. The analysis of backward citations can identify basic patents and the time that has passed since a basic innovation.

At the end of a technology cycle, the time span between major patents and current patents should be very large so they will not be discovered in the time frame

Table 7.4 Forward patent citations for combustion furnaces, 1985–2001

	Combustion furnaces						
	1985	1988	1990	1994	1998	2000	2001
Number of patents	90	77	79	71	89	89	71
Citations 0–4 years	24	16	23	17	19*	3*	0*
Citations 0–14 years	63	48*	45*	22*	19*	3*	0*
Average number of citations 0–4 years	0.27	0.21	0.29	0.24	0.21*	0.03*	0.00*
Average number of citations 0–14 years	0.70	0.62*	0.57*	0.31*	0.21*	0.03*	0.00*

* Incomplete data.
Source: European Patent Office.

Table 7.5 Backward citations for burners, 1985–2001

	Burner						
	1985	1988	1990	1994	1998	2000	2001
Number of patents	28	43	57	59	95	94	95
Citations 0–14 years	26*	25*	30*	107	152	76	62
Citations 0–4 years	26	15	13	44	65	37	27
Average number of citations 0–14 years	0.93*	0.58*	0.53*	1.81	1.60	0.81	0.65
Average number of citations 0–4 years	0.93	0.35	0.23	0.75	0.68	0.39	0.28
Average age of citations 0–14 years	3.5*	5.0*	5.8*	5.8	5.8	5.3	5.8
Average age of citations 0–4 years	2.2	2.8	2.4	2.4	2.7	2.4	2.9

* Incomplete data.
Source: European Patent Office.

of our analysis. The average number of citations in the range 0–4 years declined from 1985 to 2001, interrupted by a small peak in 1994 which can be explained by the development activities in burners (Flox burner; regenerative burner in the early nineties). However, the overall citation rate in the analysed period is rather small and the analysis should have been extended to include years before 1985 since all the indicators analysed so far have pointed towards furnaces being a mature technology at the end of the technology cycle. It can therefore be assumed that important patent activities will probably have taken place much earlier.

7.2.2.2 Scientific Publications

Scientific publications are a further indicator for the position of a technology in the technology cycle. Figure 7.10 shows the number of publications between 1974 and 2004 related to different aspects of furnaces.

In principle, the analysis of the scientific publications backs the results of the patent dynamics' analysis albeit with a slightly more dynamic development in publications. Once again, this is due to the activity in only a few fields. The field of optimisation is the main driver with a rapid increase starting in the late 1980s, which can easily be identified as connected to the availability of relevant computer capacities and their performance. The next major field is that of burners which have already been identified as a driver of the whole technology and which has even higher numbers if burners and nitrous oxides are combined. Since the beginning of the 1990s, there has also been increasing interest in heat loss, heat transfer and refractory materials. This may be one explanation for why the number of patents on refractory materials/heat insulation increased after 2000. Typically, scientific publications

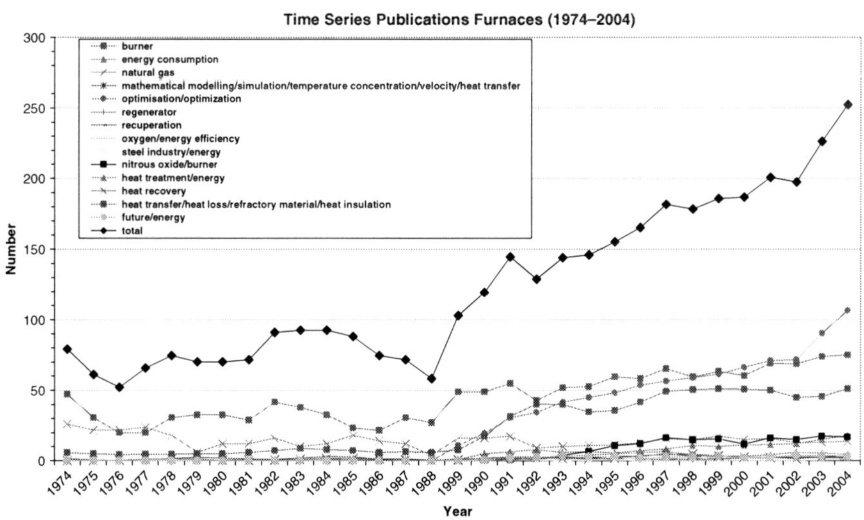

Fig. 7.10 Scientific publications related to furnaces, 1974–2004

predate patents and represent the first basic results, though not fully evaluated with regard to their economic importance.

Figure 7.11 highlights the area where the total number of publications is much smaller. Again, three issues of increasing importance can be identified during the last 15 years. One, the NO_x emissions, has already been discussed; the second – energy consumption – will remain a strong issue for furnaces due to the high impact of energy costs on furnace operation. The third smaller but still significant aspect is the field of mathematical modelling and simulation. One driver for the increasing number of modelling activities has been predicting NO_x emissions without having to build a full scale test plant. Due to improvements in computer performance and more sophisticated flow, heat transfer and reaction models, many improvements on the burner side have been possible in a fast and cost effective way.

Table 7.6 gives an overview of the different indicators used to identify the position of industrial furnaces on the diffusion curve. As can be seen in the table, most of the indicators do not clearly point to the phase on the diffusion curve, as the values of a single indicator may refer to more than one phase. Only if the overall picture given by all indicators is viewed as a whole can a robust result be obtained. However, the necessary data are not available for all indicators (columns in bold italics), or cannot be determined quantitatively. Therefore some scoring had to be made based on the interviews and other sources of information (columns in bold). When summing up the number of marked fields (italics and bold), the highest total score is obtained for the diffusion phase.

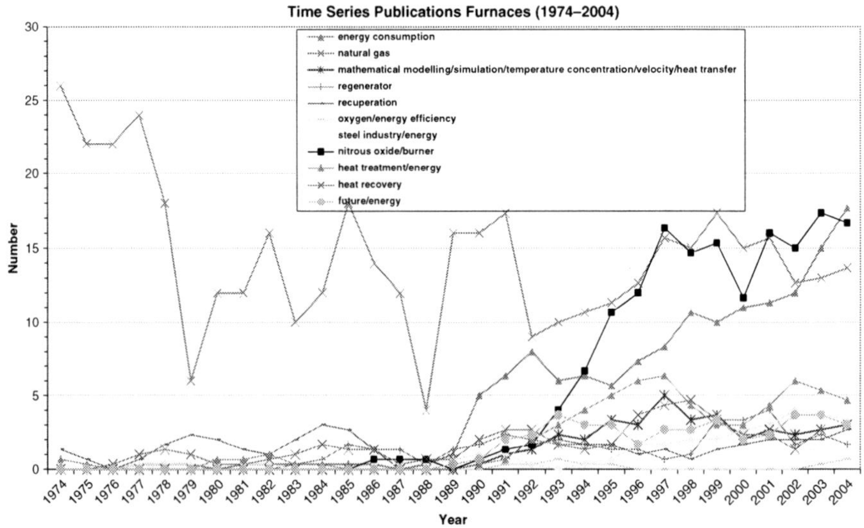

Fig. 7.11 Scientific publications related to furnaces, 1974–2004 (taken from previous figure)

Table 7.6 Set of indicators for furnaces and kilns

			1	2	3	4	5	6
Phase			Discovery	Euphoria	Disillusionment	Reorientation	Rise	Diffusion
Patents and publications								
A.1	Number of publications	n	low	*medium*	*medium*	*medium*	high	*medium*
A.2	Publication trend	n/a	*increasing*	*increasing*	stagnating	stagnating	*increasing*	stagnating on high level
A.3	Average number of citations from publications	n	***very high***	***high***	***medium***	***medium***	***medium***	***medium***
A.4	Publication trend in popular literature*	n	low	***increasing***	***stagnating***	***decreasing***	***increasing***	***stagnating***
A.5	Number of patents	n	*low*	high	*low*	medium	high	very high
A.6	Applicants: Enterprise size		major enterprises	*SME*	major enterprises	major enterprises	major enterprises n	*SME*
A.7	Applicants:		research institute	*industry*	*industry*	research inst./ind.	*industry*	research inst./ind
A.8	Trend of patents	n/a	increasing	increasing	decreasing	*stagnating*	increasing	stagnating on high level
A.9	Average number of citations from patents	n	***very high***	***high***	***medium***	***high***	***low***	***low***
A.10	Age of citations in patents	a	***low***	***low***	***medium***	***medium***	***high***	***very high***
Markets								
B.1	Market penetration	%	low	low	medium	medium	high	*very high*
B.2	Cumulated number of users	n	0	low	low	low	medium	*very high*
B.3	Cumulated number of sold units	n	0	low	low	low	medium	*very high*
B.4	Number of competitive techniques*		**low**	medium/high	**low**	1	**low**	high
B.5	Use of brand names		**no**	yes	**no**	yes	yes	yes
B.6	Share of value added*	%	***low***	***low***	***medium***	***medium***	***high***	***very high***
B.7	Number of suppliers	n	0	**medium**	low	very low	**medium**	high
B.8	Number of start-ups*	n	0	medium	low	**very low**	medium	high
B.9	Insolvencies	n	0	0	many	many	**a few**	medium
B.10	Number of outfitters	n	***0***	***medium***	***low***	***very low***	***medium***	***high***
B.11	Number of component suppliers	n	0	medium	low	very low	medium	**high**
B.12	Number of specialised suppliers	%	0	low	0	very low	low	**medium**
B.13	Type of co-operations		***development partnership***	***development partnership***	***horizontal co-operation***	***vertical co-operation***	***product partnership***	***service/processes***

Table 7.6 (continued)

			1	2	3	4	5	6
Phase			Discovery	Euphoria	Disillusionment	Reorientation	Rise	Diffusion
Frame conditions								
C.1	R+D expenditure	Euro	low	high	**medium**	**medium**	high	**medium**
C.2	Trend of R+D expenditure	Euro/a	increasing	increasing	decreasing	**stagnating**	increasing	**stagnating**
C.3	Research promotion	Euro	medium	very high	medium	very low	low	**low**
C.4	Number of proposals for R+D	n	**low**	high	very high	**low**	**low**	medium
C.5	Classification of agents		research institute	industry	industry	**research inst./ind.**	industry	**research inst./ind**
C.6	Investment in technology development	Euro	low	high	**medium**	**medium**	high	**medium**
C.7	Consideration in technological impact assessment	n	**low**	medium	medium	**low**	high	**low**
C.8	Market studies	n	***0***	***many, broad***	***medium***	***medium***	***high***	***medium***

Labelling
Indicator determined quantitatively In italics
Indicator determined qualitatively In bold
Indicator could not be determined In bolditalics

7.3 The Innovation System of Industrial Furnaces

7.3.1 The Overall System

At first sight, the innovation system of industrial furnaces seems rather complex and is characterized by a diversity of actors. There are two reasons behind this: first, the field of thermo-process technologies/industrial furnaces consists of several areas of applications and, second, the technology is not a closed one but rather a combination of different technologies (e.g. refractory, burner, automation etc.). We therefore have to analyse the furnace as a system applied in a diversity of areas and its components as partially developed and researched in scientific fields only indirectly linked to thermo-processes. As a result, the sheer variety of furnaces is abundantly clear if one considers size, capacity, material heated, quality requirements, fuel used, industrial sector involved etc. We have therefore limited our analysis to those furnaces used in the metal working industry, which not only comprise the largest share of all furnaces but are also the most important energy consumers next to furnaces used in the consumer goods industry. Thus, they are a good starting point for energy efficiency considerations.

The innovation system of German manufacturers of industrial furnaces and their suppliers shows a special characteristic compared to other industrial sectors. Apart from a few areas, German businesses lead the world market for manufacturing both furnaces and the required components with a share ranging from 20% up to 30%. However, this market share is now declining in the face of rising competition from China, India or – within Europe – Italy and, to some degree, Romania. The German export share of above 70% is also declining, but neither rapidly nor significantly.

The actors in the field are well established due to the traditional German strengths in developing and building heavy equipment. This is also true for the corresponding research institutes so that the whole system is stable with almost no new companies or research actors apart from changes caused by takeovers, insolvencies or other economic based developments.

Figure 7.12 shows an overview of the German innovation system for furnaces. The different actor groups are then analysed in more detail.

7.3.2 The Manufacturers/Industry

Looking at the manufacturing side of the innovation system, it is obvious that most companies are small or medium-sized. According to the fragmentation of application areas – and therefore the variety of furnaces – even the bigger companies tend to split into smaller or medium-sized units operating on different, independent markets with differing demands and technological requirements and/or standards. Only 10% of the German companies listed[55] are active in the production of industrial furnaces

[55] According to the database Creditreform.

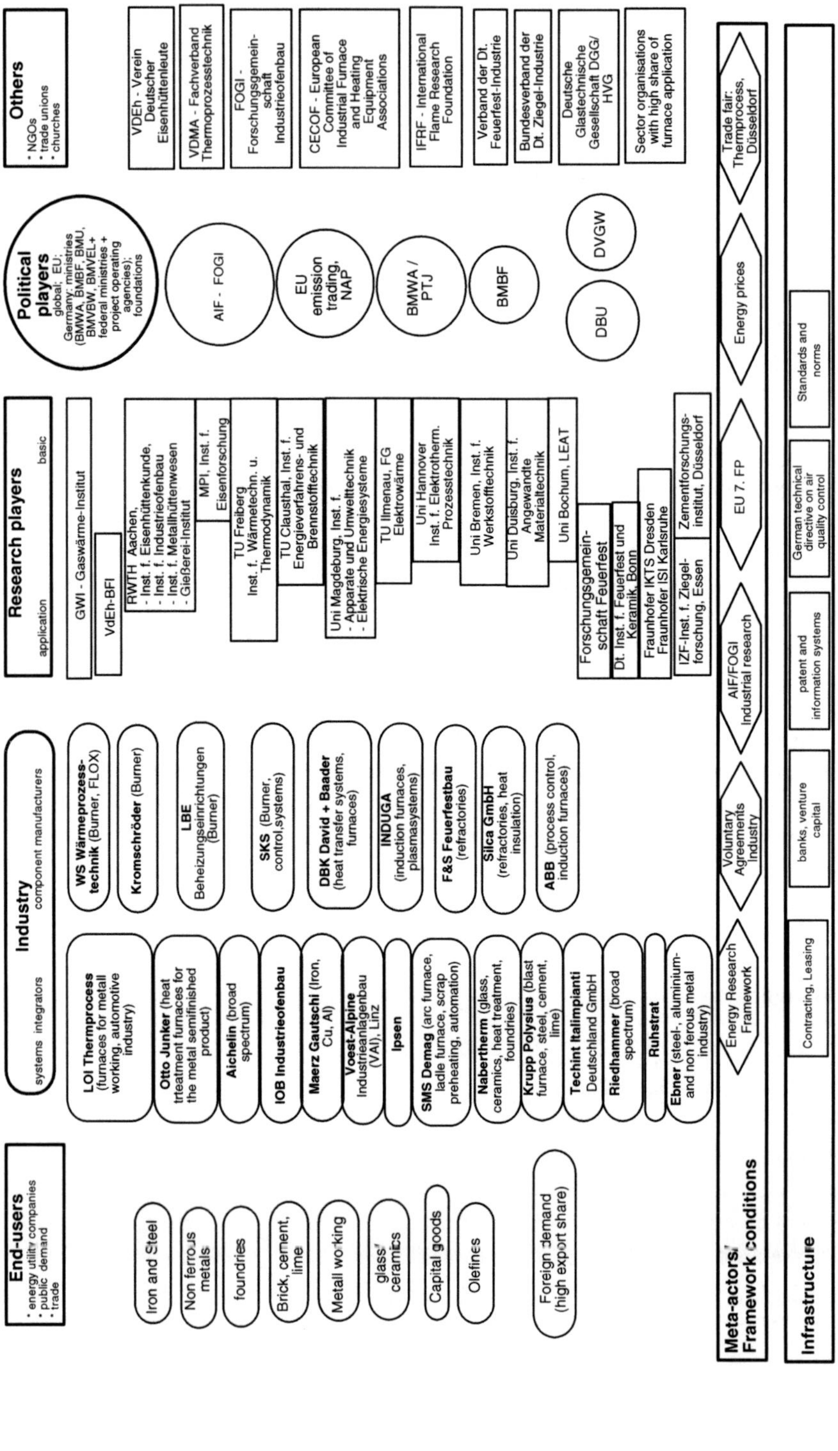

Fig. 7.12 The German innovation system for industrial furnaces, around 2005
Source: Fh-ISI.

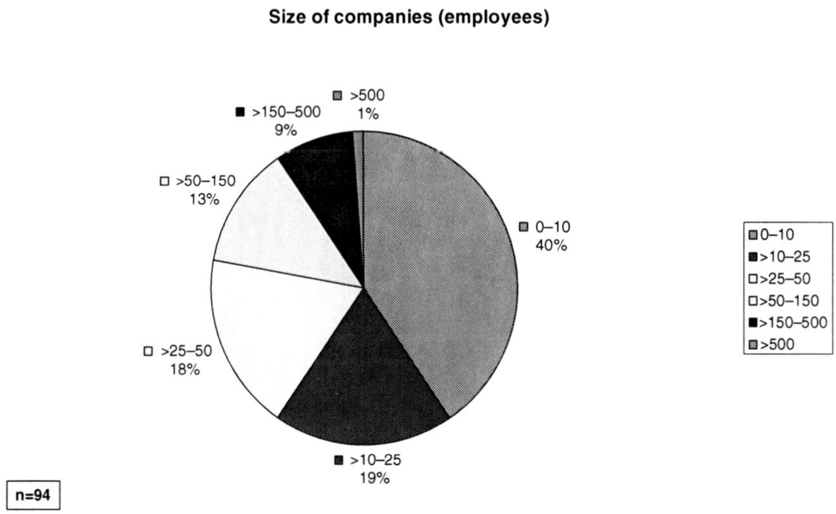

Fig. 7.13 Size of German companies by number of employees, 2005
Source: Creditreform.

and kilns and have more than 150 employees; a majority of 59% have fewer than 25 employees (see Fig. 7.13). However, these figures should not be overestimated since the database uses descriptions of the actual activities of the companies and not an impartial classification of their products.

As is the case with the overall structure of the innovation system, there is little fluctuation in the number of businesses concerned with industrial furnaces. There have been a few mergers and takeovers in the past, sometimes by foreign companies, but this is very infrequent and no trend or current development can be determined. Nevertheless, some interviewees point out that even infrequent or rare mergers harbour a potential threat to the German innovation system. On the one hand, despite clear German leadership in terms of production and sales, the business is internationalised to a very high degree. As a result, mergers or changes in the ownership of German companies often involve foreign investors. Their understanding of the conditions of the German system might be restricted if indeed they are at all interested in the national innovation system. This is by no means a normative statement but refers to the fact that investors by definition seek economic success and therefore do not necessarily pay intention to questions of innovations. This is even more valid for the business field of industrial furnaces since this is a traditional industry characterised by incremental innovation. If foreign investors own the entire company or hold the majority of a company's shares, this would have the effect of diminishing Germany's importance on the world market. A second effect of merging companies is the formation of larger enterprises or so-called associated companies leading to a difficult situation with regard to the public funding of R&D. This will be discussed later in this chapter.

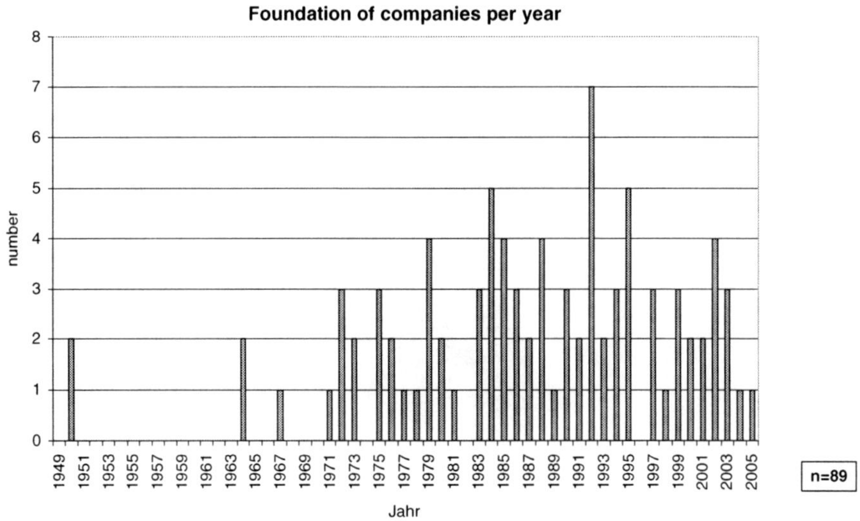

Fig. 7.14 Entrance of German companies to the market, 1949–2005
Source: Creditreform.

However, as the interviews reveal, the whole sector is still characterised by a high degree of stability and the existence of several economic niches. As we mentioned before, the following chart demonstrates that this sector is not static in every sense in terms of entrances of companies to the market. The late 1980s and the early 1990s saw a relative large number of market entrances (see Fig. 7.14). However, one has to keep in mind that nearly 70% of all companies are subsidiary companies. This supports the estimation that there is a trend towards outsourcing or merging on the one hand (the bigger companies tend to split up into smaller units or keep the name of a purchased company for several reasons, e.g. image or brands) and rather limited access for new companies on the other hand.

As stated above, German manufacturers lead the world market and therefore the changes – if there are any – are linked to the challenge presented by new competitors. On the one hand, there is growing competition from Eastern Europe and, more importantly, China, but this is limited to areas with rather low-tech qualities. Here it is unclear which role these competitors will play in the future, except that they will continue to gain market power and maybe start to work on more high-tech furnaces or components (which are by definition quite advanced in general). Of greater significance is the fact that not one of our interviewees was able to make a reasonable guess about when these changes might occur. This indicates either that German companies might be underestimating the potential challenge, or that the division of the markets involved is so stable that no real competition will result from this development.

On the other hand, countries like Italy, France, the United States and primarily Japan, compete with German manufacturers in the high-tech niches and sectors of industrial furnaces, i.e. markets with a high degree of German involvement and thus significance for German manufacturers. Their importance in terms of market shares varies in different areas of application to a significant extent, e.g. Italy dominates foundry furnaces. Nevertheless, in general, only a small market share has been relinquished by German companies to their rivals, e.g. Italian companies gained 6% in the volume of trade worldwide from 1985 to 1995, while Germany lost 2.5%. These figures differ from sector to sector, and from furnace to furnace to a certain degree but still tend to confirm the general situation.

To sum up, the situation of German manufacturers is very similar to the one experienced in many industrial sectors in the late 1970s. A very stable system of established companies is producing a mature technology and leading, if not dominating, the world market with its products. Competition is from other high-wage countries and from certain emerging economies in the low-tech sectors. The latter are not seen as a real threat to the system yet but as something worth monitoring. In addition, the system is scattered corresponding to the areas of application.

Inter-industrial cooperation is underdeveloped compared to more innovative sectors. Except for the traditional forms of cooperation in corporatist arrangements, business associations etc., the companies tend not to cooperate which constitutes a problem for the public funding of R&D as we will see later. At the same time, vertical cooperation (in contrast to horizontal, i.e. cooperation along the value chain) is by no means as useful or profitable as it might be in other industries. German companies tend to work within well structured parts of the market, relying on long-standing relations with suppliers and customers. This refers to the appraisement of our interviewees that most German enterprises act almost exclusively as business rivals. Networking therefore does not seem to be a requirement for successful operations. Companies engaged in the same or adjacent markets tend to avoid competition by dividing the market between them (e.g. for bigger and smaller constructions, or one for cement and one for the ceramic industry etc.).

Nevertheless, the manufacturing businesses are seen as the main drivers of the technology. Neither the demand side nor the research system has sufficient knowledge and/or resources to push developments. Due to the fact that this technology has been in use for a long time, the knowledge about needs and obstacles is located within the companies.

7.3.3 Public Research

Following the argumentation that there are as many areas of applications for industrial furnaces as one can think of, it is not surprising that the actors in public funded research are just as diverse and manifold as the companies involved. Even though our analysis limited these to the metal working industry and the corresponding research organisations, it still has to cope with a large number of actors.

Basically, public research is structured according to two main principles. The first is the principle underlying the whole German research system: the division into university and non-university institutes on the one hand and institutes for basic and applied research on the other. Depending on the degree of maturity of the technology, the main actors are either institutes engaged in applied research or technical universities. The latter are the most important group followed by non-university institutes and single-issue organisations working on solutions for clearly defined industrial areas like the Brick and Tile Research Institute. This organisation reveals the second main principle, that of a thematic division of labour. Some institutes deal exclusively with refractories, others with all aspects of thermodynamics.

Since industrial furnaces are a mature technology (i.e. there is no "real" or pure basic research on furnaces as a "system", although there is certainly some basic research in combustion processes or new materials which could contribute to this technology in the near future), nearly all the institutes are involved in applied and partially in basic research. However, the basic research relevant to industrial furnaces is often not performed in these institutes, but instead in institutes focusing on boilers and combustors for power generation, in areas where the total turnover in industry is much higher than for furnaces. This was already apparent from the analysis of the top patent holders, who often come from the power generation sector.

Only the Max Planck Institute for Iron Research works exclusively by assignment in basic research. All other public non-university research institutes are of course highly dependent on public and private funding. Due to the stable structure of the industrial system, the cooperation partners for public research institutes do not tend to vary that much. The relationship between the institutes and their industrial partners is characterized by established and long running co-operations. Interviews reveal that researchers are anticipating a shift in the distribution of funding. Although it is not clear where this growing competition is coming from, some stated that a fixed or even declining amount of available funding would have to be distributed among a growing number of institutes in the future. However this declining amount could not be proved based on the analysis of the data available in the federal funding database. Even if the variation of available funding is large, the trend in the recent past has still been upwards (see Fig. 7.15).

However, this feeling might be based on the fact that the basic funding of the universities is too small, forcing them to apply more and more for third party financing but not being very successful in doing so. Figure 7.16 shows that the share of universities in the available funding does seem to have declined over the past few years.

The question also needs to be followed up why interviewees from science stated that the number of actors is actually growing since one would expect a stable or even declining number following the drop in funding. One reason might be that a cutback in the public money available is forcing institutes who have been working in scientific niches so far to leave these and try and participate in the more general funding.

It should also be noted that the share of funding going to industry is very high, pointing to a very well developed product, where incremental innovations are mainly taking place in industry.

It should however be kept in mind that besides the federal funding of projects, there is additional funding available for furnace research. Alongside direct project

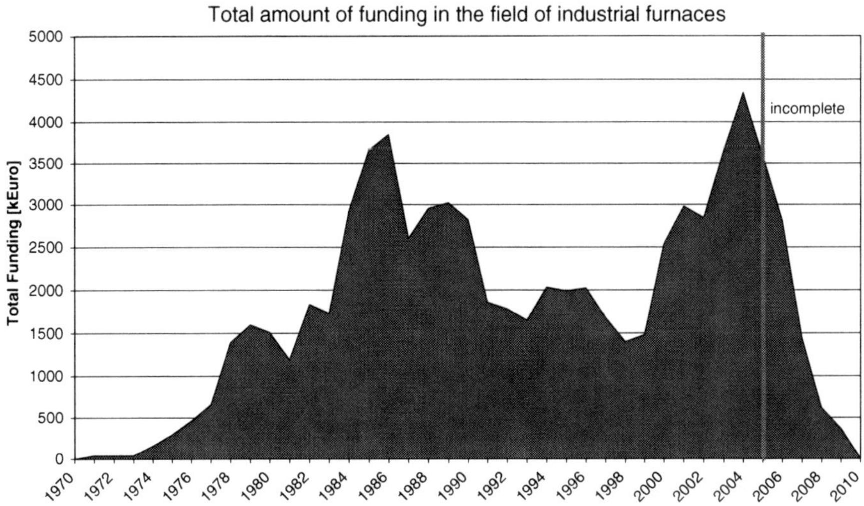

Fig. 7.15 Available direct public funding for furnaces from federal ministries in Germany, 1970–2010. Source: Förderdatenbank des BMWi, 2005

funding, the Federal Ministry of Economics and Technology also supports industrial research via the institutional funding of the AIF. The members of the AIF are research organisations in a broad field of activities. Figure 7.17 shows the main furnace related institutions represented in the AIF. The AIF distributes funds to the research organisations which in turn often pass on the funding to their research

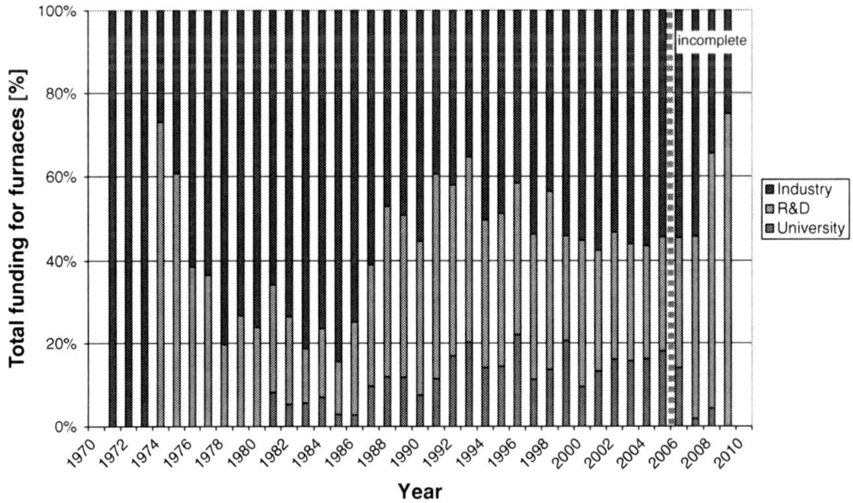

Fig. 7.16 Share of industry, R&D institutions and universities in federal project funding, 1972–2020. Source: Förderdatenbank des BMWi, 2005

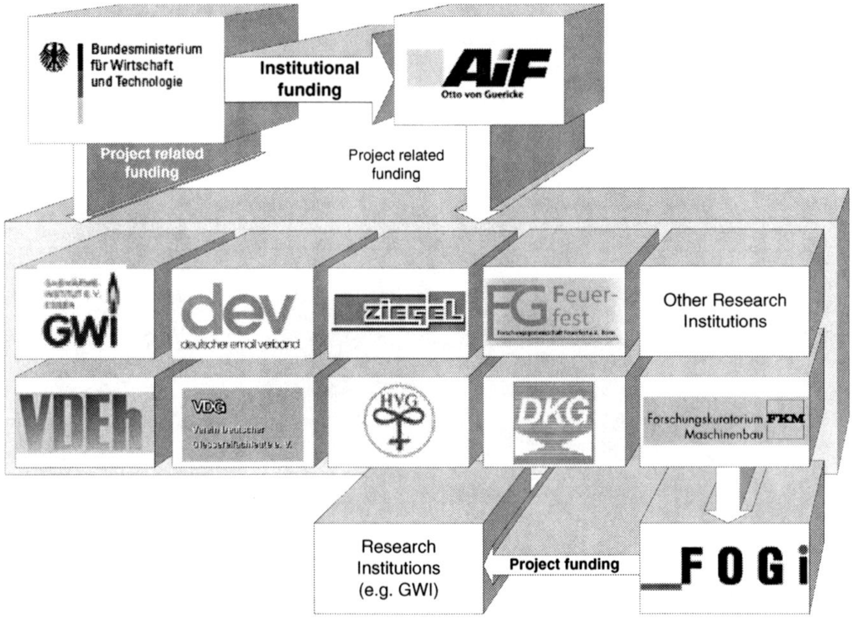

Fig. 7.17 Funding mechanisms in place and research institutes for industrial furnaces, 2005

institutions. These finally assign the funding to research projects realised in research institutes. So an institution like the GWI could receive direct project funding from the ministry, institutional funding for its own research and project-based funding from institutions who have received institutional funding themselves. Funding resources are therefore very diffuse and nobody is able to give an overview about the amount of funding spent on work related to furnaces.

The analysis of the federal funding for furnaces is therefore incomplete and a wrong impression might be formed from looking at federal funding only.

As mentioned above, the tendency to cooperate is not very well developed on the industrial side of the innovation system, which is due to the maturity of the technology and the relatively low demand for innovation. Not surprisingly, this is also true for the research system. Furthermore, from the perspective of the interviewees, there is almost no relevant competition and therefore no cooperation partner outside Germany, either in terms of industry or public/private research. Moreover, there is hardly any opportunity to cooperate with institutes abroad, since there is no conference or similar event with sufficient relevance for R&D in industrial furnaces (there is one conference on Industrial Furnaces and Boilers which takes place every two years in Portugal but with a strong focus on boilers). Almost every project is a close cooperation between a research institute and a company, i.e. most of the scientific work is focused on individual companies' problems leading to individual solutions. The different institutes tend not to cooperate with each other, which might be connected to the lack of funding/very limited amount of funding dedicated to industrial furnaces. The field is characterized by low funds with a declining tendency, so it is natural that the amount of money available for networking is small, too, or even insignificant. Some of this might change due to the research framework programmes of the EU preferring larger networks working on larger projects. However, they may even be counterproductive, since the projects will be too large for some smaller institutes if not for all. Administrative complexity might outweigh the positive effects for a scientific field working in small research projects based on cooperation with individual companies.

This situation varies for different areas. For instance, in the field of electric arc furnaces, European cooperation clearly outweighs the contribution of the German innovation system. Additionally, European competition and cooperation is growing in importance due to the European framework programmes.

7.3.4 Interest Organisations

As a result of the corporatist structure of the German political system, there is one central organisation which accounts for the majority of companies, representing 80 enterprises and 70% of the German turnover made with furnaces and/or components: the VDMA (German Engineering Federation – and, within this, the section for thermo-processing and waste treatment technology). The VDMA is active as an interest group on both the national and European level. It seeks to influence regulations regarding all aspects of the construction and application of thermo-processing installations. To pursue its goals, the association works closely with the European Committee of Industrial Furnace and Heating Equipment Associations (CECOF) in which VDMA holds the secretariat.

When analysing the German innovation system, it is even more important to mention the Research Association for the Thermal Process Industry (FOGI = Forschungsgemeinschaft Industrieofenbau) in which 25% of the VDMA members are organised. This association provides significant funding for R&D in the areas of efficiency, security, product quality, process reliability, materials and human-machine interaction: the annual budget of FOGI is about 700,000–800,000 €, about 50% of which is from public sources (AIF) and the other half from the FOGI member companies. This is equal to approx. 30% of the entire industrial research budget in the area of furnaces and helps to finance about 6–7 projects each year with an average duration of 1–3 years. Therefore, the VDMA is not just an important player within the system; it is responsible for a considerable share of medium to long-term industrial research projects.

7.3.5 Demand

As already stated, the world market is the relevant one for the German innovation system in industrial furnaces since German manufacturers lead this market in most sectors or have a significant share in areas which they do not dominate. As the

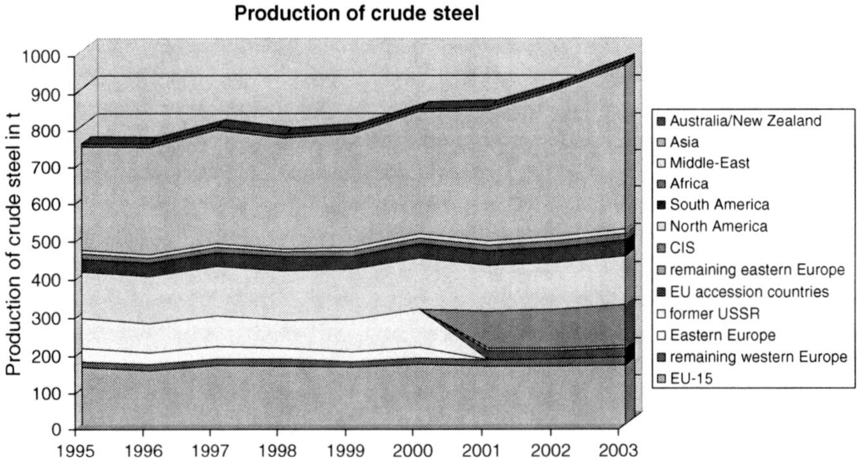

Fig. 7.18 Production of crude steel, World, 1995–2003
Source: Steelinstitute VDEH.

German system of innovation is very stable and the maturity of the technology is clear, it is not surprising that the world market is not prone to quick or significant changes. Chinese and Eastern European competitors have taken over the low-tech segments (see Fig. 7.18) and will surely try to move into more high-tech niches, but the overall structure is expected to remain stable. At the same time, the growing Chinese and Indian demand for steel, steel products and other high quality goods of the metal working industry is driving the markets. As long as the increased demand for steel products is large enough to extend the total market for furnaces or furnace components, the absolute size of German industry will probably not change, even if the total market share is declining. These two counterbalancing effects actually produce a very stable situation. However, it is very uncertain whether the Asian demand for steel and therefore furnaces will continue in the future, or, as some experts recently predicted, whether a peak in demand might be reached in a few years followed by a subsequent decline. The producers of iron and steel have to anticipate future developments in the market and align their demand for furnaces and kilns correspondingly. Industrial organisations and associations are already warning against the possible outcome of overproduction followed by a drop in demand. However, as the following two charts reveal, the trend for the world's production and for Germany's import and export of iron and steel is unbroken (see Fig. 7.18 and Fig. 7.19 respectively).

The demand side is characterised by a strong conservative attitude towards innovation. This is not only because well-established large companies dominate the sectors demanding industrial furnaces. In addition, the technical processes guaranteeing the quality of the product, which form the focus of the companies' work, are highly sensitive and susceptible to malfunctions. As typically high costs are involved if the required product quality is not achieved (which is due to large equipment sizes and high throughput), innovations will not take place if possible benefits are limited and

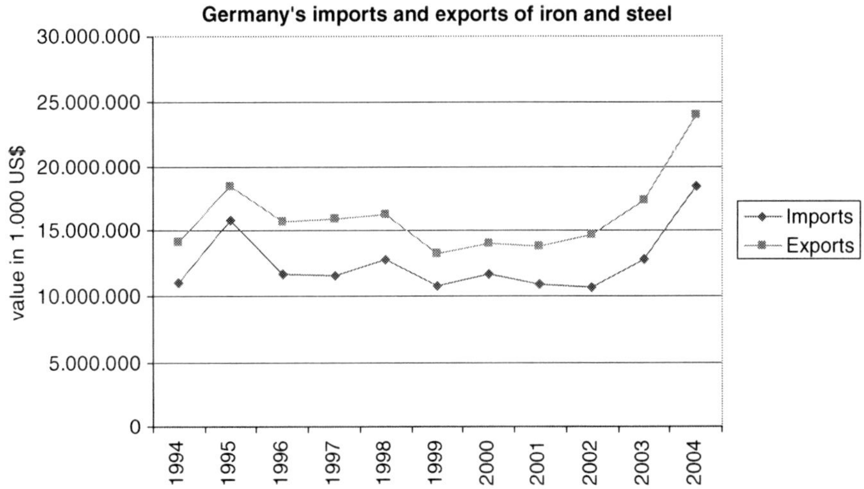

Fig. 7.19 Germany's iron and steel imports and exports, 1994–2004
Source: FTS.

not all the effects known in advance. Enterprises generally try to avoid the uncertainty which is a natural part of innovation. Nevertheless, the industry is anticipating the re-fitting of older constructions with more modern refractories, automation, thermal insulation and preheating units. Furthermore, it is expected that users will partially replace their burners. Changes to the furnace auxiliaries, which can be dismantled in the case of problems with the new equipment, will have the highest chances of being taken up by the industry.

Again, the diversity of industrial furnaces and application areas makes it impossible to give a realistic assessment of either the current market or its development in the future. This is also true for any statement about the actual division of the market, upcoming demand or other issues. Interviewees stated that there is not one reliable market study available with regard to volumes, sales figures and current tendencies. The demand is – in general – characterised by the fact that there is no alternative technology able to replace furnaces.

7.3.6 Public Policy

Overall, public policy does not seem to play an important role in industrial furnaces, certainly not compared to other energy-relevant industrial sectors. Looking at the expenditure of the Federal Ministry of Education and Research (BMBF), it is quite clear that the importance of basic research – which forms the focus of the BMBF – experienced a rapid decrease expressed in the total funding of industrial furnace technology from 1985 on – after a period of even quicker growth in BMBF-funding following the energy price crises in the 1970s. The total volume of the support programmes for furnaces fell from 7 million € in 1985 to approx. 125,000 € in 2000. Because of the weak demand for basic research and the corresponding concentration on applied research, the BMBF research on energy is driven by industrial demands; this development ultimately reflects the priorities of the industry. Above all, it is striking that there is a certain amount of funding available but no major support programme in place.

The role played by the Ministry of Economics (BMWi) is somewhat different. It traditionally promotes competition, the market and the competitiveness of German SMEs and provides a significant amount of money to foster innovations developed in SMEs. Support for R&D in the field of thermo-process technologies – especially burners and waste heat utilisation – formed part of focus areas if the Government's 5th Programme for Energy Research (2005, p. 37), the R&D programme is open to any valuable idea and not limited by deadlines. Manufacturers also benefit from the subsidies granted via the non-specific research and innovation programmes supporting medium-sized businesses administrated by the AiF.

Nevertheless, direct or indirect public funding which is predominantly provided by national ministries seems to be essential for industrial research (mostly done by SMEs) and the relevant supporting research institutions. Industrial research is, in turn, the precondition for the continued improvement of the technology and for maintaining German companies' international success.

Technological enhancements have to be prepaid which represents a high economic risk for SMEs in general and in particular in a sector characterized by very pronounced economic swings.

Jobs in public research institutes are highly dependent on public research funding. If the industrial subsidies were not supplemented by public funds, approx. two thirds of the jobs in research institutes would cease to exist. In terms of basic research, the DFG is the main public actor but does not support the scientific area of thermo-process technologies with a "Sonderforschungsbereich".[56]

The EU is becoming increasingly important as a catalyst and as the pertinent political level with regard to standards and regulations. Here it is striking that the research projects funded by the EU lack a thematic main focus and that the expenditures as well as the number of projects vary widely over time. This not only affects the continuity of non-industrial research, but could intensify the problem that companies independently decide on research orientations of the whole field and may neglect interesting approaches without the impulses/correctives provided by publicly funded science.

There is another "soft" reason for the rather low involvement of public policies. In the public's view as well as in politics, industrial furnaces are seen as symbols of outmoded, mature, heavy industries. As the interviews indicated, they are often linked with images of smoking chimneys. As a result, it does not seem appropriate to channel money, especially taxpayers' money, into seemingly outdated industries, even though this ignores the high importance of the sector for German industry.

Public policy regulations have an influence on the sector especially where energy consumption is affected. Emission trading is seen as one of the most important

---

[56] Collaborative Research Centre.

political developments, at least for the furnace operators of a certain size, but more concrete expectations have not yet been expressed. As many steel companies with their large furnaces are affected by emission trading, energy efficiency has received additional attention. Especially at a time when production capacities are used at a rate of nearly 100%, the emission certificates allocated at no cost will not be sufficient. Therefore companies need to decide whether they are going to purchase additional certificates or identify and realise energy efficiency measures in order to fulfil their emission obligations.

7.4 Implications for Public Policy

7.4.1 Bottlenecks and Opportunities

The maturity of the technology gives rise to a situation where no genuine technological bottlenecks are hampering development. The importance of constant process and product qualities together with the risk aversion of the technology users tend to hinder breakthrough innovations and foster incremental innovations anyway.

As both the interviews and the patent/publication analysis revealed, there are at least two or three interesting developments that could contribute to improved furnace technologies. A promising energy efficiency approach concerns *fuel consumption*, i.e. the use of more efficient burners and superior insulation. A bottleneck for the latter is the need to substitute materials that are suspected of causing cancer. Furthermore, it is now highly unlikely that any breakthrough in R&D could significantly lower the total consumption of fossil fuel or energy. Basic research on the fundamental *exploration of combustion processes* might yield knowledge in the future that could lead to radical changes in the technology.

Economically, the industry is well positioned as stated above. The main bottleneck is caused by the risk aversion of the users of industrial furnaces which is both an attitude and a consequence of the technology itself. However, this should be seen against the industrial background – as one interviewee stated – the willingness to take risks is actually sufficient. Furthermore, the manufacturers of furnaces are mainly SMEs which experience the usual problems of not being able to fund their own R&D. It remains unclear whether additional financial support for R&D would initiate any change because SMEs also have only a limited number of staff to handle R&D projects. Although we have stated here that there is no current or foreseeable bottleneck which could be overcome by more research, the ability to keep abreast of technology developments is crucial for companies to maintain their economic success. The mergers within the industrial part of the innovation system or the growing importance of (foreign) investors do create a bottleneck that could threaten or is already threatening the system in terms of public funding for R&D (as a basis for innovations in SME). According to the strict EU definition of SMEs being beneficiaries of public funding – which has to be followed by the German funding agencies – some might lose their eligibility to apply for funding. If we nevertheless assume that SMEs are the drivers of innovation and that foreign investors or owners may not invest in R&D in Germany, such a development could lead to a gap between the role SMEs are expected to play regarding innovation and the role they are able to play in the future. Associated enterprises above a certain threshold of either number of employees or amount of turnover will not receive public funding for their R&D. Assuming that the trends towards economic concentration continue or even increase, the German innovation system could then face serious problems as indicated above.

As mentioned before in this report, the main bottleneck in the political system is the failure to see industrial furnaces as a high-tech industry. Public funding is available but no specific support programme with a decent budget has been implemented recently. This harbours the risks of first, missing the opportunity to give impulses for the orientation of research and second, not being able to bundle financial and other resources to try to overcome the obstacles (like the replacement of carcinogenic refractories). This is not to say that a larger budget, more bundled funding or the implementation of specific support programmes would guarantee a change within the system since there is no clearly identified bottleneck. It seems to be more important to guarantee continuous funding over longer periods, even if this implies a lower annual budget. Research costs are often determined by personnel costs and fluctuating funding does not help research institutions to hold on to highly educated and trained staff with the result that new projects often mean starting from scratch with regard to building up the necessary knowledge. In addition, the decrease of relevant funding especially in basic research might cut off Germany's companies from important and promising developments. This may ultimately bring about the loss of technological and economic leadership on the world markets. This could occur despite the lack of obvious bottlenecks. The fact that no-one expects or can foresee any breakthrough innovations does not mean such developments are impossible. At least in some areas basic research is exploring problems which may be very relevant for industrial furnaces. Therefore, it could be decisive to keep up with these proceedings even if they are not directly linked to thermo processes and their industrial application at present.

One area in which it would be beneficial to bundle financial resources is the research done in power generation (boilers, combustors, modelling). The transfer of research results from power generation to furnaces is still taking too long or is not taking place at all. One possible solution could be to foster closer cooperation between these two research fields in joint projects.

This could also help to overcome one of the main deficiencies in the system, the weak networking/cooperation between companies and research institutes. This is, of course, one outcome of the degree of stability attained; companies do not rely upon cooperation to gather sufficient resources for innovation. The maturity of the technology results in a situation characterized by competition where there are well-established supplier-customer relations and client bases and little necessity for networking. Organisations responsible for the administration of support programmes/projects complain that this can hamper the development of more radical innovations which require companies to bundle their resources. Although all those interviewed ruled out breakthrough innovations in the near future, it is not clear whether innovations beyond the usual incremental progress could be achieved faster (or at all) if the lock-in in terms of non-cooperation were overcome.

7.4.2 Policy Recommendations

The field of industrial furnaces/thermo-process technologies is far too complex and diverse to be able to provide simple answers and/or recommendations. Even focusing on only one of the areas of application – e.g. the metal working industry in this report – does not significantly reduce the variety of problems and possible solutions. Industrial furnaces are mostly larger systems consisting of very different components. Some of the latter are all-purpose techniques; some are specially designed, utilised or even constitute unique parts of furnaces. Nevertheless, it is possible to put forward some rather general recommendations.

The first and most important recommendation is that public policy should increase its efforts in fostering innovations in general. As some of the components and much of the knowledge used and required in the construction of (enhanced) furnaces are derived from research only indirectly linked to furnaces (if at all), supporting such projects is essential. Specific research on the topic of combustion in general should be especially fostered and – even more importantly –technology transfer via the organisations or support programmes for such projects has to be improved considerably.

These recommendations indicate two different directions policy could take. One refers to research slanted towards industrial – i.e. applied research. As the interviews showed, this is and will remain the crucial focus for this technology, but even though no breakthrough innovations are expected, it would be very reprehensible to allow German researchers and manufacturers to lose touch with developments that might form the basis for a breakthrough in the future. Basic research is a precondition for industry to succeed and therefore has to be supported. Of course, public policy cannot protect the innovation system from all potential threats. Based on our interviews, we therefore recommend the *improved participation of the scientific community* – i.e. institutes and chairs working on thermo-processes etc. – *in decision processes concerning the development of support programmes* and thematic focuses as one possible first step. Moreover, *the isolation of German research in thermo-processes must be overcome.* Although this isolation is based on a scientific and economic advantage compared with other countries and a stable share of patents and turnover on the world market, it could become a threat to Germany's competitiveness in the long run. As economies grow outside Europe, an increasing number of R&D activities will start in other regions of the world, spawning new competitors and new ideas and making it necessary to strengthen international cooperation between industry and research. To achieve this, a change in attitude will be necessary. Industrial furnaces are an important part of the German economy in total and constitute a high-tech sector which needs permanent R&D. Furthermore, the lack of cooperation within the German innovation system regarding furnace technologies, identified as a systemic bottleneck, could be overcome if national policies addressed the need for more or less institutionalised cooperation between German enterprises, research institutes and other key players. To achieve this, it would be necessary to fund network infrastructure using cooperative R&D projects and to increase funding in interdisciplinary research groups in order to link furnace technology research to R&D in other interesting fields. Besides such measures, it is vital to strengthen the links between German and international public and industrial R&D via joint research projects and an increased openness of national funding and support for actors from abroad.

The other direction concerns questions of how to deal with industrial furnaces in order to achieve greater efficiency and lower energy consumption. As a system made up of numerous diverse components, industrial furnaces have many set screws. At the same time, this does not mean simply reducing the amount of energy provided to the processes themselves. Derived from the patent and publication analysis, we would like to stress the *importance of the permanent improvement of burners*. The success of the FLOX- and regenerator burner shows what is possible. To achieve this, combustion processes have to be explored still further which reinforces the argument made above that basic research still has a lot to contribute to this issue. Furthermore, additional issues should be addressed which might also be beneficial for many other industrial sectors. Major challenges still lie in the development of sensors to analyse gas composition, temperature and flow velocity in harsh environments such as those in furnaces. The analysis of the actual conditions in a furnace is a prerequisite for exact furnace control and therefore for producing high quality products. Furthermore, there is also an energy impact here as energy wastages due to necessary security margins could then be reduced or eliminated.

Earlier in this section, we already indicated the gap which may develop between the – generally held – role of SMEs in innovation and their chances of obtaining public funding to fulfil this role. To some extent this problem is based on the EU's definition of eligible SMEs, which is not likely to change, and even if it did, national policies would still have to be coordinated first. National policies should be aware of this potential gap and its related problems.

As energy prices will continue to rise in the future, new and improved approaches will be necessary to reduce the energy consumption of furnaces. However high temperature processes will always require heat at high temperature levels and will release heat at lower temperature levels, so there should be a greater focus on optimising the use of waste heat. At present, any waste heat recovered is typically used in the plant itself (e.g. combustion air preheating) or on site (e.g. steam generation) which limits the amount of heat that can be reused. Iron and steel companies typically have a heat surplus which they cannot fully exploit so that cooperations should be promoted which use the available waste heat from furnaces off site (e.g. Alu Norf, Düsseldorf), or use the surplus waste heat for power generation via steam or ORC cycles.

Material research, which might be driven by the power generation industry, will also play a key role in the future. Closer attention will be paid to gasification technologies (see Chap. 6) in the future which may help to advance the search for better refractory and insulation material. In general, public policy has to address technology transfer from the increasingly important field of new materials to the industrial furnace sector. Although it might not be clear from the start which innovations in the material sciences may contribute to improving furnace technologies, it would be a major failing to overlook such applications.

Continuous monitoring should be made of material substitution and changing production chains. As production technology advances, some furnace applications may no longer be required (e.g. thin slap casting made some reheating furnaces superfluous). Heat treatment may also become unnecessary, if changes are made to the material shaping processes or if other materials are used for the products. If steel products were replaced by plastic or glass products, this would obviously have an impact on the furnace industry.

References

Baake, E. (1995). *Möglichkeiten der Energieeinsparung und Emissionsreduzierung bei industriellen Prozesswärmeverfahren (Possibilities of energy saving and emission reduction by industrial process heat processes).* ETG-Fachberichte, Vol. 57, S. 93–98.

Ban, S. D. (1993). *Dramatic energy savings through gas research.* Industrial Heating, Nr. 8, S. 4–6.

Barth, G., Groß, U., & Wulf, R. (2004). Wärmeleitfähigkeit keramischer Fasermatten – vergleichende Messungen mit unterschiedlichen Methoden. Keramische Zeitschrift, Nr. 9/10, S. 554–561.

Beneke, F. (2006). VDMA: Personal Communication, Frankfurt.

BFI. (1991). *Energieeinsparung an Wärmeöfen durch Zustellung mit neuzeitlichen Faser-Dämmstoffen.* Förderkennzeichen Nr. 03E8172A/9.

BINE. (Juli 1996). *Gasdurchströmte Strahlungswände in Industrieöfen.* BINE Projekt Info-Service, Nr. 4, FIZ Karlsruhe.

Bradke u. a. (1996). *Energieeffizienz, Strukturwandel und Produktionsentwicklung der deutschen Industrie.* IKARUS, Instrumente für Klimagas-Reduktionsstrategien. Abschlußbericht Teilprojekt 6 – Industrie. Monographien des Forschungszentrums Jülich, Band 19, Jülich.

Brunklaus, J. H., & Stepanek, F. J. (1994). *Industrieöfen: Bau und Betrieb. Brennstoff- und elektrisch-beheizte Öfen – Brennerkonstruktionen (Industrial furnaces – construction and operation. Fossil-fuel and electric furnaces burner concepts).* Esen: Vulkan Verlag, S. 800.

CADDET. (2003). High-performance industrial furnace using high temperature air combustion – application to a heating furnace, IEA, Paris.

Calon, M., Dahm, B., Klima, R., & Stroo, D. (2002). Energieeinsparung und Qualitätsverbesserung durch Regelung der Verbrennungsgasatmosphäre in Warmöfen. Stahl und Eisen, Nr.2, S.75–80.

Collier, D., Smirnov, V., & Stroup, S. (2004). Entwicklung des Hochgeschwindigkeitsbrenners "ThermJet" mit reduzierter NOx-Bildung. Gaswärme International, Nr.3, S. 139–143.

DIN. 24201: *Industrieöfen, Wärmöfen und Wärmebehandlungsöfen (Begriffe) (1982).* Ausgabe Mai 1982, Berlin: Beuth Verlag.

Dög, A., Walter, G., & Krause, H. (2002). Minimierung der Wärmeverluste in den Industrieöfen Gaswärme International, S. 353–358.

Fleischmann, B. (1997). *Konventionell beheizte Glasschmelzöfen für die Hohl- und Flachglasherstellung im deutschsprachigen Raum – Ergebnisse einer HVG-Umfrage – Teil 2. Glasqualität und Energieverbrauch.* Glastech. Ber., Vol. 70, Nr. 3, N27–N34.

Foreign trade statistics, US Census Bureau Washington D.C.

Gaswärme Institut. (1989). *Untersuchung und Optimierung der Strömungsverhältnisse in Öfen beim Betrieb mit Rekuperatorbrennern.* Förderkennzeichen Nr. 03E8522A/7, Gaswärme Institut Essen.

Hammer, H. (1989). *Konzeption und Betriebserfahrungen mit dem neuen 200-t/h-Hubbalkenofen einer Mittelbandstraße.* Stahl und Eisen, Bd. 109, Nr. 22, S. 73–81.

Heckenthaler, T., Röder, M., & Klima, R. (1996). *Energieeinsparung an Wärmöfen durch modellgestützte Überwachung und Steuerung der Brennmedienverhältnisse.* Europäische Gemeinschaft für Kohle und Stahl; Vertrags-Nr. 7210-EB/108,

Klima, R., Sucker, D., Voigt, H., & Wick, D. (1998). *Verbesserung der Wärmeübertragung in Walzwerksöfen insbesondere durch Weiterentwicklung und betriebliche Erprobung zusätzlicher adiabatischer Hilfsheizwände.* Bericht EUR 16040 der Europäische Kommission Reihe: Technische Forschung Stahl.

Krupp Stahl, A. G. (1991). *Einsatz von Rekuperatorbrennern zur Abwärmenutzung an einem Wärmebehandlungsofen mit hoher Prozesstemperatur.* Förderkennzeichen Nr. 03E8522B/0, Krupp Stahl.

Muhle, H. (1997). *Wie gefährlich sind Faserdämmstoffe?* Fraunhofer Mediendienst, Nr. 4, Thema 3.

Müller, R., & Pietsch, J. (1991). *Untersuchungen zum Energienebenverbrauch an Anlagen (Schmelz-, Wärmeöfen, Umformeaggregate) in der Metallurgie.* Blech Rohre Profile, Vol. 38, Nr. 5, S. 387–390.

Münch, F. T., Raffauf, F., Röder, M., & Klima, R. (1996). *Untersuchungen zur Verbesserung der Energieausnutzung und der Temperaturgleichmäßigkeit.* Stahl und Eisen, Vol. 116, Nr. 3, S. 49–54.

Munko, A., Jäger, F., Köhne, H. (2002). Neue Brennertechnik für Strahlheizrohre in Thermoprozessanlagen BWK, Nr.9, S.1–16.

N. N. (2004). Entwicklung und Optimierung von Oxy-Fuel Rekuperatorbrennern. Gaswärme International, Nr. 2, S. 81–83.

N. N. (2004). Forschungs- und Entwicklungsvorhaben: Themenschwerpunkt "Industrielle Gasverwendung". Gaswärme International, No. 2, S. 78–80.

Radgen, P., Jochem, E. (Hrsg.). (1998). *Zwischenbilanz zur rationellen Energienutzung bei Thermoprozeßanlagen, insbesondere Industrieöfen.* Abschlußbericht, Karlsruhe: Fraunhofer ISI.

Radgen, P. (Hrsg.). (1997). *Dokumentation des Fachgesprächs Elektro- und Kombiwärmeverfahren.* Düsseldorf, 9. Oktober 1997, FhG-ISI, Karlsruhe.

Radgen, P. (Hrsg.). (1997). *Dokumentation des Fachgesprächs Feuerfest.* Bonn, 24. Juni 1997, FhG-ISI, Karlsruhe.

Radgen, P. (Hrsg.). (1997). *Dokumentation des Fachgesprächs Ofensteuerung und Ofenregelung.* Düsseldorf, 17. April 1997, FhG-ISI, Karlsruhe.

Radgen, P. (Hrsg.). (1996, November). *Dokumentation des Fachgesprächs Brennersimulation.* Düsseldorf, 24. Oktober 1996, FhG-ISI, Karlsruhe.

Rüdiger, J., Brüser, G., Ruhe, H. (2003). Modernisierung und Optimierung von Stoßöfen in einem Grobblechwalzwerk. Stahl und Eisen, Nr.5, S. 37–41.

Statistisches Bundesamt: *Fachserie 7, Reihe 7 (Außenhandel nach Ländern und Gütergruppen).* Wiesbaden, Ausgaben 1986 bis 1994

Stoller (1996, November 1). *Feuerfeste Faser fangen sich Kritik ein.* VDI Nachrichten, Nr. 44, S. 40.

Stoller (1997, September 5). *Steinwolle: Ohne Krebsverdacht.* VDI Nachrichten, Nr. 36, S. 26.

VDMA. *Statistisches Jahrbuch für den Maschinenbau.* Frankfurt, Verschiedene Ausgaben.

VDMA-Einheitsblätter 24202. (1980). *Industrieöfen – Klassifikation.* Ausgabe Januar 1080, Beuth Verlag, Berlin.

Walter, G., Krause, H., Dög, A. (2003). Wärmedämmung des Abgasführungsrohres von Rekuperatorbrennern – eine Nebensache? Gaswärme International, Nr.5, S. 276–280.

Wick, D., Voigt, H., Klima, R., Sucker, D. (1994). *Verbesserung von Wärmeübertragung und Energieausnutzung in Walzwerksöfen – Erprobung von Hochemissionscoatings und adiabaten Hilfsheizwänden*. Gaswärme International, Vol. 43, Heft 7/8, S. 331–343.

WS Wärmeprozesstechnik. (1994). *Flammlose Oxidation. Vermeidung der thermischen Stickoxidbildung durch Verbrennung mit hoher interner Abgas-/Luftvermischung*. Förderkennzeichen Nr. 0326939A, WS Wärmeprozesstechnik.

Chapter 8
Conclusions and Recommendations

The *EduaR&D* assessment aimed at developing a methodology to support decision makers in energy research policy in their choice of which technologies to support (*what*) and the form this support should take (*how*). The assessment applied a complex concept combining three different analytical approaches (energy economics analysis, technology cycle analysis, and innovation systems analysis) for the first time to four distinct energy technologies. In doing so, it significantly sharpened the underlying methods and optimised their combination. Furthermore, the study generated experience on the prerequisites, added value and potential limits of the concept and derived a whole range of recommendations regarding its further improvement and future application.

This chapter reflects on the conclusions of this study, the observations made in the four applications of the suggested methodology and the results of the international workshop held on February 21–22, 2006, with regard to the general methodology and the technology-specific conclusions which may be generalised to some extent.

It has to be stressed that the *EduaR&D approach chosen in this project* should be regarded as

- *an aid to reflect on the current portfolio of energy research* at the national or supra national level (EU) or at the company level of a large company with several business areas in the energy field, or as
- *an assessment and policy design tool at the meso level* of energy technology areas and technology priorities.

It should *not* be seen, however, as an aid for funding decisions about concrete, individual research projects. The suggested approach does not replace basic political R&D decisions such as e.g. whether to conduct applied research and development in the car industry or to prolong research funding for high risk options such as energy fusion. The assessment is designed to be applied under specific circumstances (e.g. comparable risks involved) and specific conditions:

- after basic decisions have been made on how to split the research budget into institutional and project funding (e.g. relating to the budgets among the BMWi, BMU, and BMBF at the national level),

E. Jochem (ed.), *Improving the Efficiency of R&D and the Market Diffusion of Energy Technologies*, DOI 10.1007/978-3-7908-2154-3_8, © Physica-Verlag Heidelberg 2009

– after certain politically based decisions have been made at the macro-level on how to split the project-based R&D budget (e.g. renewables versus efficient use of energy versus carbon capture and storage).

The assessment approach suggested here relates to those research areas and budgets which are project-oriented. The suggested approach is also useful when decisions have to be made regarding changes in the R&D portfolio of highly funded energy technology areas on the one hand and less well funded areas on the other.

The authors have become increasingly convinced that, at this level, the additional analytical steps are worth doing in the case of new technologies with a high future economic potential in the field of energy (which may involve annual investments and energy savings of several billion Euro at the national level and even more at the global level). The four technology examples chosen in this study have this economic significance and may be used to test and evaluate the additional benefits stemming from the process suggested here.

However, it must also be emphasised that this study only developed the methodology of the assessment; the concept still has to be applied in its full spectrum after completion of this study. One cannot expect to see full fledged efficiency improvements in R&D and market penetration as a result of this project, but only as a result of its application in the future. It has to be understood as an accompanying process to improve policy-decision making regarding the R&D and market introduction processes of future energy technologies.

The conclusions (which try to keep a critical distance to the results achieved) are structured in a methodological part (Sects. 8.1 and 8.2) and a technology-oriented part (Sect. 8.3). Implicitly, the conclusions suggest a whole set of recommendations that have mostly been discussed within the chapters on the individual technologies (Chaps. 4, 5, 6, and 7). Finally, Sect. 8.4 focuses on the essential results of the analyses and looks at further activities to improve the efficiency of R&D in the field of energy.

8.1 The Overall Approach and the Interaction Between the Three Assessment Areas

The major characteristic of the ISI-EduaR&D concept is that it is a combination of three distinct analytical approaches: energy technology and economics, technology cycle analysis and innovation systems analysis. Each of these conceptual approaches is a challenge in its own right and has its own specific value. However, it is their combination that generates the added value of the ISI-EduaR&D project. The major lessons learned from the interplay of the three approaches and from their individual application within ISI-EduaR&D can be summarised as follows:

- *Methodological fit*: the design and the combination of the three approaches must fit the technological area considered and its identified elements. In-depth knowledge of both the technology and the context in the three areas energy technology and economics, technology cycle, innovation systems is indispensable prior to designing the empirical work. This prerequisite may not be easy to meet in the immediate future within existing energy research administrations given the interdisciplinary character of this concept. But such interdisciplinary assessment teams can be developed over time and supported by the manual developed in this project (see Annex 9.1). The results of this study clearly advocate the establishment of such interdisciplinary teams.

- *Procedure of the analysis:* the (partial) assessments made by each of the three approaches should be carried out fairly simultaneously to ensure efficiency and to enable cross-fertilization between the different methodological standpoints (see also Fig. 8.1).
- *Analytical interaction:* structures and target-oriented and quality-assuring routines must be set up in order to secure the necessary feedback loops between the three approaches and the analysts and to secure the precision needed for the identification and assessment of the next steps in R&D, market introduction or market diffusion policies. Ideally, a process manager with basic capabilities in each of the three analytical approaches should be nominated for this task. Such moderated interaction helps

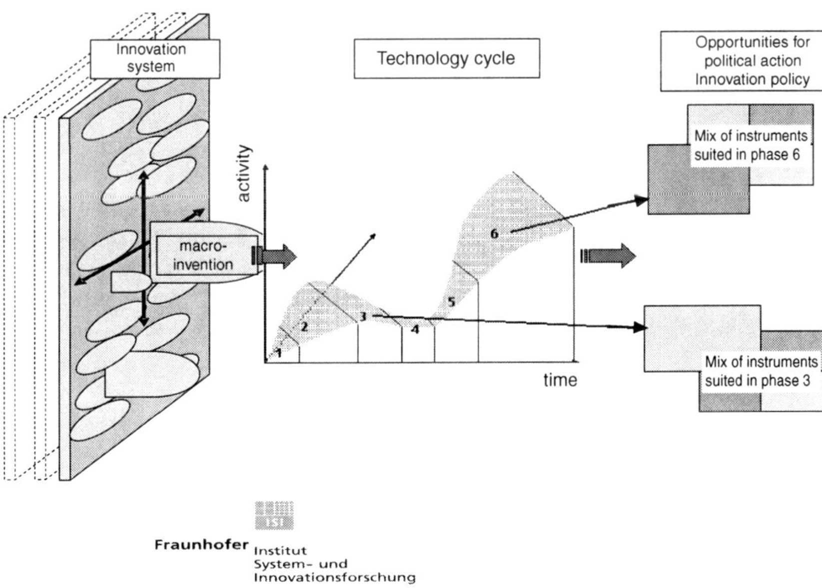

Fig. 8.1 Innovation system, technology cycle, and policy options and their interplay during the simultaneous three pillared analysis
Source: Fh-ISI, 2006

  - to define – as early as possible within the process – the *level of technology and important technical elements,* particularly those representing the critical steps in the R&D process ("bottlenecks" e.g. acceptable operating hours before maintenance becomes unavoidable) or the market introduction process (e.g. acceptable cost or price levels);
  - to inform about and assess the present and future advantages/disadvantages of *complementary or competing technologies* as these may support or hinder the market introduction of the new energy technology in question (e.g. the PEMFC has to compete in stationary applications against the internal combustion engine or cogeneration powered by Stirling engines and the MCFC or the larger SOFC against micro-turbines or large cogeneration plants combined with heat generating boilers; see also Sect. 8.3);
  - to *constantly check the quality of interim results* from the three methodological areas (e.g. problems with the identification of patents and citations, or estimates of future cost and technological performance) and adjust methods and search strategies accordingly (e.g. start by interviewing experts first because of the difficulties in identifying relevant patent classes, journals (or even proceedings) and cost data);
  - to define and *identify the relevant actors within the innovation system*, both to analyse the performance of the technological system and to obtain the expert support needed for the various methodological steps. The analysis may start with patent or bibliometric analysis if key actors have limited knowledge or with interviews if key actors are well informed or patent class identification is difficult.

- Due to the interdisciplinary approach, it is of great importance to define a *common terminology* among *all the analysts* contributing to the study as well as among the *supported decision-makers.*
- The authors of this report compiled a *manual on how to most effectively apply the proposed assessment concept.* It is expected that this manual will help to keep the efforts and resources required for an analytical endeavour of the suggested type at an acceptable level, especially in light of the additional benefits which are expected during the R&D phase and market penetration. The major rationale behind this is that bottlenecks and market obstacles will be identified very early on and the role of competing technologies will be analysed with greater intensity and objectivity.
- Moreover, how a technology fits into the existing or future export profile and thus the underlying economic structures of the country is one overarching variable for the funding decision. The EduaR&D assessment approach can also assist in determining how the specific technologies analysed relate to the existing technological strengths of the country in the context of global technology competition.
- The assessment process suggested here and discussed at the international workshop in February 2006 runs the risk of *requiring too many resources to quantify all the indicators* suggested (e.g. patent and citation development, anticipated

costs at the stage of market introduction, estimated cost decreasing potentials by learning and economies of scale, present market shares). The question of whether the additional efforts relative to the present decision making process will generate sufficient additional benefits in terms of greater R&D efficiency or whether they would result in higher administrative net costs and lengthier decision processes cannot be answered generally but has to be considered case by case. Certainly, an extensive and hence lengthier decision process can only be accepted if the efficiency gains of the assessment are expected to offset the delay due to the pressures of the international technology race.

8.2 Conclusions Derived from Observations in the Three Conceptual Areas

Besides the general observations mentioned above, other methodology- or technology-specific observations are summarised the following sections.

8.2.1 Energy Technology and Economics Analysis

- An objective assessment of the economic potential of a new energy-related technology depends on *many assumptions about future framework conditions* such as energy price levels and relative prices, climate change emission targets, and technological breakthroughs. Further assumptions are anticipated cost reductions which may also depend on exports. Existing bottlenecks, public acceptance of the new technology, and the market development of the competing technology also reflect assumptions on boundary conditions. This leads to "weak potentials" depending on the set of assumptions made for assessing the economic potential of the new technology. Because of these uncertainties, the proponents of a new energy technology tend to overestimate favourable boundary conditions and the stakeholders in the conventional or alternative solution tend to do the opposite.
- Most of the new technologies emerging from R&D have *traditional (or even simultaneously emerging) competitors*; these constitute technical as well economic benchmarks representing minimum R&D targets with regard to technical characteristics and cost (e.g. the fuel cell and its traditional competitors: internal combustion engine driven cogeneration, micro-turbines, Stirling engines). In some cases there may also be a technological race between new technologies (e.g. the three technical options of carbon capture from large fossil fuel conversion plants, nuclear power or renewable energies with large electricity generation potentials). In these cases, a very high quality standard for the analysis would call for a similarly intensive analysis of the traditional or simultaneously competing technology. In practice, however, this may prove too costly, too time consuming and/or extremely comprehensive in many cases.

- *New technologies often cost much more than their traditional competitors* at market introduction. This is due to a variety of reasons such as low production series, the unfamiliarity of the new technology, lock-in effects or high switching costs. The assessment of learning and economy of scale effects becomes crucial as the assumptions about leapfrogging investments (and the risks involved) (e.g. vacuum insulation, PEM fuel cell) involve uncertain projected future developments of the relevant boundary conditions such as energy prices and favourable/unfavourable policies.

This step is vital for the R&D assessment process as, on the one hand, the proponents of new technologies tend to overestimate cost reduction potentials and oversee any further cost reduction potentials of the competing technology. On the other hand, those with a stake in the traditional energy technology tend to overemphasise the present high costs of the new technology at its market introduction.

8.2.2 Indicator-Based Technology Cycle Analysis

- The nature of a technology, its position in the technology cycle – assessed early on by experts – and its related/anticipated markets determine the selection and definition of indicators that can be used in a meaningful way. A thorough selection and definition process is required; there is *no mechanistic application of indicators.* This implies that particularly the comparison of technological options may not be straightforward (e.g. industrial furnaces in their market diffusion position compared with CO_2 capture from fossil fuel power plants which may be essential in the decades to come in order to meet the future CO_2 targets of electricity demand).
- A two-stage approach is needed for the *publication and patent analysis* whereby the key word search for the analysis is systematically derived from a first thorough analysis of the scientific and popular literature in order to prepare the search strategy for the final analysis. Obviously, it must be borne in mind that not all scientific progress in selected technologies is reported in scientifically reviewed journals, e.g. passive houses and buildings are mostly reported in German or Scandinavian construction and architectural journals or in the proceedings of European conferences; the technical progress of industrial furnaces is mostly described in non-scientific journals of the associations of furnace manufacturers or furnace users. As a result, a multi-step approach may be required for certain technologies to identify the relevant publications.
- Each *analysis of a technology cycle requires a tailor-made selection,* not only of indicators, but also of the database used to retrieve data for indicator construction. For example, there may be technology-specific publication traditions in non-reviewed journals (e.g. industrial kilns). In some cases, a technology is mainly developed in or confined to a non-English speaking set of countries and is not a typical candidate for a global market product. In these cases, there is

the risk of an under-representation of these technologies in databases that are predominantly based on reviewed journals or on the English language or which have a strong international focus. In this case, it could be necessary to search for additional or alternative publication databases in order to assure that the research strategy fits the specific features of each technology.

- If technologies are in the *phase of very applied research or even incremental development during market penetration, their technical progress is often not described in scientific journals*, but in expert journals or proceedings. Interview partners contacted within the innovation systems analysis may hint at these facts and the citation analyst has to be aware of this possibility.
- Apart from this issue, the various publication databases as well as their corresponding journals or conference papers also differ in their coverage of specific fields of research. If the analysis, for instance, deals with a *young, just emerging technology*, there may not be many established journals in this field. So the main scientific discussion may take place in all kinds of grey literature. The selection of databases must be adjusted to these circumstances. It becomes apparent that it is crucial to gain all this information about the technology to be analysed at a very early stage within the assessment process. As mentioned above, interviewing experts might be one way to gather this information.
- To generate or refine the *keywords and phrases* for the patent and publication analysis, the systemic examination of standard literature, articles, or abstracts via a *text mining* tool could be useful. For example, this might be necessary if a technology does not commonly use particular technical terms. As described in the chapter on passive houses and buildings (see Chap. 5), the most frequently used terms and phrases were extracted from some scientific articles using a text mining tool. Subsequently, the information obtained forms the starting point for further patent or publication analysis.
- The *definition of a key technology* via *key words and patent classes* used to define the patent search strategy is an extremely *challenging process for complex, modular or systemic technologies*, but is essential for meaningful research:

    - In many cases, the *existing patent classification does not reflect the new technology in clear search words*, as this must – by definition – lag behind technology developments. This means that the relevant patent classes have to be identified by very experienced patent analysts supported by interviews with the R&D staff and product managers of technology producers and with relevant researchers in the science system. One way to tackle the problem of defining the appropriate patent classification is to start with a set of actors who can be expected to have applied for a patent in a specific technology that is meant to be used as such, rather than being used in different technological contexts. The link to the relevant patent classes can then be derived by checking the applications of these specific actors. This has to be done in order to most efficiently link the relevant energy-related products and technologies to the existing patent classification.

- In some cases, particularly in technological areas with incremental innovations (e.g. industrial furnaces), innovations may take place in apparently unrelated fields of technology such as specific materials which are not at all obvious to the general R&D analyst. The definition of key technologies and relation to patent classification may, therefore, require *multiple feedback loops between patent analysts and technological energy experts in industry and academia to check the plausibility* of the data collected and the indicator results, to assist in interpreting the indicators (e.g. what stage of the technology cycle) and in identifying technical or economic bottlenecks.
- As every technology is usually comprised of more than one element, *several search strategies in parallel* may be useful for the set of components of a given technology or system as interim results may influence the next steps of the analysis.
- Even if the technological area, the patent class or citation search word are all clearly identifiable, the number of patents or citations of publications may be difficult to interpret *if the technology considered can have different fields of application* (e.g. improved sensor and control techniques or improved qualities of materials). This *dual or multifunctional use of a new technology* does not make it easy to differentiate whether the particular patent is specifically devoted to energy applications or whether it can be used for other applications as well. The dual (or multiple) use of a new technology (or an element of it) may result in overestimating the significance of the number of applications within a specific patent class.
- This situation can usually be managed by defining specific search words for the expected energy application and conducting search runs over the texts of the identified patents of dual use.

- *A Trade Mark* analysis was tested to complement the market diffusion analysis. A clear link of a technology to a trademark classification is often not possible (exceptions: "Passivhaus" in Germany and "Minergie P" in Switzerland for the passive houses and buildings), but trade mark searches – also of components of the new technology – may be used as a first step to define the relevant actor set, and it helps to determine whether the key market players of a given technology are represented and what kinds of trade marks they hold.
- *Market indicators* of market introduction and market penetration *such as* cost reduction potentials of the assessed technology and its competitors, the present and projected market shares or international competitiveness of a particular technology as indicators were not sufficiently covered in this analysis, as this study was focussed on the development of the assessment methodology.
- In order to assess the contribution of public support for the development of energy technologies a lot of effort has to be invested for data collection of the German funding situation, especially as regards relevant institutional funding (e.g. for the institutes of the Helmholtz Society, the Max-Planck-Society, and the universities) and specific budgets like those provided by the AIF (Arbeitsgemeinschaft der industriellen Forschungsinstitute). After the introduction of the

programme-oriented research at the Helmholtz Society, this area of institutional funding is becoming much more transparent and open for an assessment developed in this study.

8.2.3 Technological Innovation Systems Analysis

The innovation systems analysis seeks to determine all components (actors and artefacts such as important norms, laws etc.) that influence and shape the generation and diffusion of relevant technological knowledge, its transfer into innovations and the diffusion of the innovation in the market. It shall lay open the division of labour and interaction of all the actors involved in order to understand how the technology is generated, diffused and exploited and who the most important actors are. Only on that basis one can discover weaknesses in the actor constellations of the technological innovation system, in the interactions between the various actors and in relevant framework and policy conditions. Furthermore the systems analysis approach enables to detect future opportunities in the shaping and application of technologies. The weaknesses and opportunities are subsequently the basis for concrete policy recommendations.

- The starting point for the analysis of the technological innovation system is to *understand the peculiarities of the technology itself*, since these peculiarities shape the mode of knowledge production and transfer, the actor constellations and interactions and the market structures at the various stages of the production and market introduction of the technology.
- The first empirical step is to identify all the relevant actors and components of the system. This identification will be based on context knowledge, document and internet analysis as well as a set of exploratory interviews. It will contain an identification of the major interactions and relations between the components of the technology innovation system.
- A graphic representation of the innovation system, its actors and their interactions has proven helpful in the interviews. It can also be used to identify bottlenecks of or barriers to the new technology and possible actors to overcome them. This figure is likely to be revised frequently, maybe even after each interview. If the innovation system considered is developing rapidly, configurations from the past (e.g. five years back) showing projections for a few years ahead may be very instructive.
- The crucial step for the analysis is an appropriate selection of interview partners. The criteria for this selection must be based on involvement/interest, representation and context knowledge. In addition, a check with international experts is recommended for technologies which are or will be traded on an international scale. The selection must ensure that representatives are included from all the important components of the innovation system, taking into account the heterogeneity of actors as well (e.g. not restricting the selection to only large companies but asking representative SMEs as well if SMEs play a significant role). The

interview should be compiled step by step over time in order to incorporate the knowledge of the interviewees and to optimise the added value of the interview process. If a technology can be clearly defined via patent or publication analysis, key actors may also be identified via numbers of patents or relevance in publication databases.

- Three waves of interviews are needed:
  - early interviews with context experts who have a bird's eye view in order to help define the relevant level of analysis and identify the most important elements of the technological level and the innovation system considered;
  - an intensive interview programme with actor-specific questionnaires, but with as much overlap as possible between the various questionnaires to identify differences in the stakeholders' views;
  - after completion of the analysis experts should be contacted again to assist in interpreting the indicator data.
- *A common understanding of the technology and a common use of the terminology* are needed between the *project partners and the interview partners* in order to avoid misunderstandings and misinterpretations (e.g. "pilot plant" was defined differently by different interview partners, as was the "techno-economic potential" of the technology considered). This implies that the interview text should be accompanied by a glossary in the annex with clear definitions of the main expressions.
- The interview programme must bear in mind that the mobilisation of the interview partners needed seems to be dependent on the phase within the technology cycle of that technology. For example, in the case of minor or incremental innovations (e.g. the industrial furnaces) there seems to be a trend that the willingness of experts to devote time to interviews is reduced compared to those interviewees who represent and promote a new energy technology.
- It might be useful to employ specialized software (MaxQDA, Atlas-ti) to sort information from the interviews on the basis of certain search questions.

8.3 Observations and Recommendations Regarding the Four Energy Technologies

Although recommendations have been given at the end of each technological chapter (Chaps. 4, 5, 6, and 7) the most relevant observations are taken up here, followed by a summary of key recommendations for each technology in the next chapter.

8.3.1 Fuel Cells

Possible future role in the energy system: many energy technologists believe the future of the three types of the fuel cell to be very bright because of their potential efficiency improvements and the low emissions associated with them (zero emissions at the point of use in case of hydrogen). The application potential is considered to be high in stationary uses (decentralised heat and electricity production) and in mobile uses due to the very large market of road transportation. Optimistic visions assume a 10–20% share of electricity generation in a future economy powered by fuel cells and even higher shares in road transport. However, these visions tend to overlook the further possible improvements of competing technologies (internal combustion engine, Stirling engine, micro-turbines), the necessary infrastructural innovations (e. g. hydrogen) and the substantial cost reduction requirements of the different types of fuel cells necessary for them to compete successfully in open markets with alternative technologies.

Position in the technology cycle: the three types of fuel cells are at somewhat different stages in the technology cycle: the MCFC is ready for market introduction but faces the typical problem of a new technology which is rather costly due to lack of economies of scale for its production and the lower cost of its technical rivals (engine driven co generation and micro-turbines). From the technical point of view, the phase of euphoria has almost passed for PEM and SOFC and further R&D activities are necessary for these two types to match the technical performance of their competitors and/or the necessary cost level for fuel cells to be technologically and economically ripe for the market. Only the DMFC has reached a standard which allows its use in niche markets like caravans or yachts despite poor efficiency levels.

Technical, economic and other bottlenecks: regarding the PEM and SOFV type of fuel cells, there should be an even greater focus of research funding on the *technological bottlenecks both in the system and in the components* (in general extending the life-time of stacks and reducing costs; especially for the PEM: i. e. improve the functions, decrease the costs of membranes and catalysts and make the system less sensitive to CO and the sulphur content of the feed).

Furthermore, the technical improvements and cost reductions of the traditional (or new) competing technologies of fuel cells have not been sufficiently considered in the past by the research community promoting fuel cells. In addition, the severity of the technological bottlenecks of the PEMFC with regard to maintenance intervals, weight per capacity, and necessary low production cost have been underestimated and some stakeholders (including politicians) are still likely to do this today. This may result in a situation in which the technology is being promoted without an in-depth understanding of the technical bottlenecks and the financial risks involved if the cost targets have to be met by large gains in the economies of scale.

8.3.1.1 Recommendations

The major recommendation in the field of fuel cells is to overcome the technology bottlenecks described above. Naïve market introduction programmes without strong R&D components might even be dysfunctional in the long run. The development of a national strategy should be strengthened, not in a uni-directional way, but in a flexible, reflexive manner accounting for the potential of competing technologies and the bottlenecks in fuel cell developments.

(1) Research should concentrate more on aspects leading to cost reductions by envisaging large production series and related production lines. This means higher priority being given to explicit technological targets in research programmes and projects (institutional funding and project funding).
(2) The discrepancy between technological bottlenecks of the PEM and SOFC and the financial risks involved in mass production on the one hand and their underestimation by stakeholders on the other also calls for *stronger links between research funding* and *application-oriented activities* and the establishment of a fuel cell strategy.
(3) Although the technology of MCFC seems to be more developed, R&D is still required to lower the costs and improve the periphery systems, which are responsible for more downtime than the stack in the ZIP project plants. The DMFC has already created a market niche in caravans or boats. R&D is necessary to improve the efficiency to broaden the field of possible applications. These could be prime candidates to test for potential market introduction funding.
(4) German policy should aim to co-ordinate a German position on fuel cell priorities in the 7th Framework Programme of the EU more intensively and visibly. This process requires the formulation of a concrete vision and technical targets for Germany which involves the major stakeholders in a transparent process. Furthermore, schemes should be developed that enable extra-European co-operation and small-scale international co-operation with multi-lateral national funding.
(5) The coordination of research and innovation funding should be improved not only between Federal ministries, but also between ministries at the level of the Bundesländer. Several actors report that other Bundesländers' programmes are not accessible to companies/institutes outside the related Bundesland, even though the company or institute could contribute to and benefit from the programme.

8.3.2 Passive Houses and Buildings

Possible future role in the energy system: Approximately one quarter of the primary energy use in Germany is required to heat buildings. This can be reduced by almost 90% using modern technologies and concepts. Saving and efficiency potentials of this magnitude (nearly 4,000 PJ by the end of this century) are not achievable in other areas. The passive house is one important option for doing so. Admittedly, it is easiest to apply passive house concepts and components in new buildings, but they also play a significant role in many areas in the energy-related refurbishment of older buildings. Even if less challenging building concepts should prevail in the long run, passive houses have an important role to play as a technology driver.

Position in the technology cycle: this technology is at the stage of market introduction (early rise) in the market segment of new detached and two-family houses. In large new apartment and office buildings, the technology is close to market introduction because some demonstration buildings have been constructed during the last few years. With regard to retrofitting existing houses and buildings, the technology is considered to be in the re-orientation phase, struggling with specific aspects such as heat loss bridges, relatively high costs and acceptance issues of building owners and renters.

Technical, economic and other bottlenecks: the technical, cost and acceptance bottlenecks are in particular:

- the integration of control functions in inexpensive solutions,
- inexpensive window systems meeting the high standards for passive houses or buildings, and
- convincing information or marketing strategies for those who believe the indoor air quality is lower if regulated by ventilation systems (compared with open window ventilation) in order to increase the acceptability of passive house design and operation.

In future, avoiding overheating in summer will become a more important task due to the changing customer configuration, increasing standards of quality and comfort and global warming.

More importantly, lack of knowledge of the new technology among architects, installers and investors in houses and buildings as well as among home and building owners (refurbishment) is increasingly an obstacle to the market diffusion of passive houses and buildings and may even be an outright threat if this lack of knowledge results in inefficient or malfunctioning passive houses or buildings (e.g. too low or high room temperatures after absences). This might even generate a backlash against the passive house development if it becomes associated with the image of an unreliable concept with a high risk of malfunctioning heating or air conditioning system (as was the case in the development of the heat pump in the 1980s).

From an economic perspective, the obstacles have two aspects:

- as a competing option with low energy houses and buildings, intensifying capital use as up-front capital to reduce and substitute energy use and long term energy costs is limited by the capital availability of the investors, but also by the lack of life cycle cost analyses by planners, architects and investors.
- The lack of knowledge and of life cycle cost analyses often give rise to the conclusion that passive houses and buildings – whether new or refurbished – are expensive and risky.

The regional limitation of the new technology, the fact that it is restricted to a few European countries like Germany, Austria, Switzerland and Scandinavia, also contributes to the small production series of technological components such as highly efficient window systems, ventilation systems equipped with heat recovery or adapted control equipment. This may change in the next decade.

8.3.2.1 Recommendations

Financing and cost issues are the crucial factor in most cases in the decision for or against a passive house. Some basic approaches exist to target the demand, which have to be employed simultaneously to be successful.

(1) The willingness and – above all – the ability to pay for the higher initial building costs have to be increased. Appropriate financing subsidies and interest reduced loans including no payback in the first years etc. should be (and are) in place to bridge the financial gap. Appropriate accounting has to be done to ensure the reduced life-cycle costs of the building are considered.
(2) Increasing the customers' ability to pay for housing, however, does not drive down the unit cost immediately. It may not even provide an incentive to do so. Public procurement of passive buildings can have two advantageous effects. First, it may increase the number of units demanded and hence induce scale effects in the supply of passive house components. Second, through a strict bidding process it may reduce the profits of passive house suppliers and their component suppliers theoretically to an almost competitive level.
(3) The risk and uncertainty about the amortization of the passive building and its technical components call for novel approaches. In combination with the suggested public procurement, one could think about new private public partnerships to launch new models of operation and financing for passive buildings. The underlying idea would be to separate ownership, financing and use of the building to enable even risk averse users to live in a highly energy efficient building such as a passive house.
(4) For passive house suppliers, the suppliers of passive house components and system integrators, financial subsidies are essential to foster the development of and the demand for passive houses.
(5) Arguing in favour of financial support for the demand side in no way rules out the option of funding the development of new technologies. Rather the opposite. The arguments given above highlight characteristics of the technologies which can contribute to the widespread diffusion of passive buildings.

Due to its high standards, the passive house approach is driving the further development of components and promoting the development of completely new approaches. Since all of this as well as the experiences with structural optimization are also used in building codes (EnEV, Energieeinsparverordnung[57]), new buildings and building refurbishments, the passive house approach is taking on an important role as technology driver which should not be underestimated.

---

[57] Energy Conservation Ordinance

8.3.3 Carbon Capture from Central Fossil Conversion Plants

Possible future role in the energy system: More than 300 Mt CO_2 per year are emitted by central energy conversion plants in Germany. This amounts to around one third of the total CO_2 emissions of this country. This share is quite representative for many countries in the world. As the CO_2 emissions have to be reduced within this century in absolute terms in spite of global economic growth and as other options like renewables and efficiency will not be sufficient to do so in large countries with enormous coal reserves (hard coal in China, India, Australia, and North America, lignite in Germany), the capture and storage of carbon is one of the major options available to meet this target.

Position in the technology cycle: All three technological concepts are still in the first rise phase ("euphoria") of the technology cycle. None of the concepts has been applied on an industrial scale so far. However there have been several announcements made about the intention to invest in CCS power plants by electricity generators and oil and gas companies which envisage different technologies. With these projects on the horizon, there is the chance that CCS technologies will take another step in the direction of market diffusion or – if larger problems occur than expected – will move onto the phase of disillusionment. In the light of this situation, the present priorities ("first, reducing energy demand for capture and secondly improving the process reliability") should be reversed in favour of the process reliability avoiding technical failures and related shrinking acceptance of the technology.

Technical, economic and other bottlenecks: in order to avoid the risk of severe disillusionment, it makes sense to concentrate research funds on problems that might otherwise result in the perceived total failure of the technologies. With respect to the pre-combustion technologies, the overall low availability of the few existing IGCC power plants makes many actors doubt the reliability of the applied system set-up. The oxyfuel technology faces a number of problems, among them the need for more efficient ways of producing the necessary oxygen than the cryogenic method currently proposed. The most critical point regarding operability is the combustion of fuels under the high temperatures of oxygen rich conditions or the alternatively envisaged recirculation of the flue gas.

Besides the principal operability and reliability of the first pilot and demonstration plants with CCS, the costs of the electricity produced should be a main focus of research policy. The cost issue is not decisive to avoid major set-backs arising from the first pilot plants. Nevertheless, the actors in the innovation systems need a clear perspective on how technology research will contribute to bring down the cost of electricity from CCS plants to the level of future market prices. In other words, the costs for capture and storage will have to reach the level of future emissions trading market prices for CO_2. The main research challenges with respect to costs lie in the reduction of the energy penalty for the capture of CO_2. Within the area of post-combustion technologies, the amine scrubbing technology needs to be improved considerably in order to reduce energy requirements and costs for the capture plant. For oxyfuel technologies, reducing energy demand for oxygen production is a core challenge with respect to the energy penalty.

The geological storage concepts are not yet fully developed and represent a critical risk to the overall realisation of the system of carbon dioxide capture and storage. Long-term tests may be necessary to gain the necessary level of certitude about their leak tightness. This refers directly to the existing geophysical methods that are not accurate enough to provide evidence about at least 99.9% of the CO_2 stored. This accuracy is necessary as leakage rates of even 0.1% of the stored mass will already result in too large cumulated losses within a few hundred years.

The geological storage concepts are not yet fully developed and thereby impose a critical risk to the overall realisation of the system of carbon dioxide capture and storage.

8.3.3.1 Recommendations

(1) Allocate significant R&D efforts to understand and operate reliably the capture processes. A reliable operation of the announced pilot plants and demonstration plants will be very important to avoid a severe disillusionment.
(2) Concentrate technical R&D activities on specific bottlenecks, e.g. inexpensive production of oxygen, safe operability of the combustion of fossil fuels under the high temperatures of oxygen rich conditions or the alternatively envisaged recirculation of the flue gas.
(3) Analyse the cost reduction potential of the three technical alternatives of carbon capture depending on the time available, capital intensity and prices of emission certificate schemes.
(4) Long-term tests are necessary to gain the necessary level of certitude about the leak tightness of geological storage. Beyond the exploration of real storage sites, sufficiently accurate monitoring techniques are required to provide evidence about the CO_2 stored.
(5) Prepare grounds for storage activities by investigating present legal and societal conditions for CO_2 capture and storage and developing solutions for a legal framework and for public outreach.

8.3.4 Industrial Furnaces

8.3.4.1 Possible Future Role in the Energy System

Although there are new developments that might limit the use of industrial furnaces, there is no sign of any overall substitution since these developments are restricted to only a few niche applications. While new materials (plastics, non-ferrous metals, wood products) could partially substitute iron and steel, it remains uncertain which areas of application will be affected. At the same time, it is not clear whether such a development would result in new types of furnaces being developed, or whether the technology would simply be reduced in total production. Overall, the world's demand for iron and steel as well as for other metals is continuing to grow, actually strengthening the importance of furnaces in the near future and the next decades.

8.3.4.2 Position in the Technology Cycle

Industrial furnaces are clearly at the diffusion end of the technology cycle. Not only is the technology itself mature, but the analysis of patents and scientific publications additionally reveals that the technology is characterised by incremental innovation and a relatively small total number of patents/publications compared with other technologies which are still in the start-up phase of the technology cycle.

This picture changes however to some extent if the focus of the analysis is shifted from the system to the *components*. Even though they are completely integrated in the production processes using furnaces, some components still show a much more dynamic development (e.g. burners, high temperature insulation material). However, because no technological breakthrough is foreseen in the near future, the components are in the diffusion phase, not one of the first phases of the technology cycle.

8.3.4.3 Technical, Economic and Other Bottlenecks

Applying more efficient burners and superior insulation materials represents one promising approach to increasing energy efficiency. A bottleneck for the latter is the need to replace fibre materials that are suspected of causing cancer.

The main operational bottleneck is the *risk aversion of users of industrial furnaces*, which is an attitude, but also a consequence of the technology use itself. Furthermore, the manufacturers of furnaces are mainly SMEs, so there are the usual problems associated with these types of firms, e.g. lack of financial resources for their own R&D. It remains unclear whether additional financial support for R&D would initiate any change, since SMEs are also typically restricted in handling R&D projects due to their limited number of staff. The developments within the industrial part of the innovation system regarding mergers or the growing importance of (foreign) investors create a bottleneck that could prove a threat, or is already threatening the system in terms of public funding for R&D (as a basis for innovations in SME). According to the strict EU definition of SMEs as the beneficiaries of public funding – which has to be complied with by the German funding agencies – some might lose their eligibility for funding. If we assume that SMEs are the drivers of innovation and that foreign investors or owners do not necessarily invest in R&D in Germany, such a development could result in a gap between the role German SMEs in this technical field are expected to fulfil regarding innovation and the role they are able to play in the future.

The main issue or bottleneck in the political system is the failure to see manufacturing of industrial furnaces as a high-tech industry. Public funding is available on a project basis, but no specific support programme with a decent budget has been implemented recently. It is clearly important to guarantee continuous institutional funding on R&D over longer periods, even if this implies a lower annual budget per year.

The fact that not one of the national experts expects or foresees any breakthrough innovation should not be interpreted as meaning such developments are impossible. At least in some areas basic research is exploring problems (such as combustion research) which may become very important for industrial furnaces. Therefore, it is decisive to keep up with these developments even if they are not currently linked directly to thermo processes and their industrial applications.

This could also help to overcome one of the main deficiencies in the system, the weak networking/cooperation between different companies and between different research institutes. In contrast, networking seems to be functioning quite well on an individual level between companies and research institutions. Organisations administrating support programmes/projects complain that this could hamper the development of more radical innovations which require companies to bundle their resources. Although all those interviewed ruled out the possibility of breakthrough innovations in the near future, it is not clear whether innovations beyond the usual incremental progress could be achieved faster (or at all) if the lock-in in terms of non-cooperation were to be overcome.

8.3.4.4 Recommendations

(1) The transfer of research results from power generation to furnaces is still taking some time or not taking place at all. One possible approach to overcome this bottleneck could be to link these two research fields more closely in joint projects.
(2) The isolation of German research on thermo processes must be overcome using joint research projects and an increased openness of national funding and support actions for actors from abroad. One way to overcome this weakness might be support for increased cooperation at the European level in the Framework Programmes.
(3) The lack of cooperation within the German innovation system regarding furnace technologies could be overcome if national policies address the *necessity of more or less institutionalised cooperation between German enterprises, research institutes and other key players*. To achieve this, it would be necessary to *fund network infrastructure using cooperative R&D projects and to increase funding in interdisciplinary research groups*.
(4) Derived from the patent and publication analysis, we would like to stress the *importance of the continuous improvement of burners*. This issue also links fossil energy use and emissions. To achieve this, combustion processes have to be explored still further. Major challenges still exist in the *development of sensors to analyse gas composition, temperature and flow velocity in harsh environments* such as those occurring in furnaces.
(5) The German government and the European Commission should reconsider the concept for SMEs and larger companies. Because of the developments in the business sector with regard to the mergers and associations of rather independent SMEs, there may be a gap developing between the – generally anticipated – role of SMEs regarding innovation and their chances of receiving public funding to fulfil this role. To some extent, this problem is based on the EU's definition of eligible SMEs and its limitation in either number of employees or size of economic turnover. Important companies for innovation might be greatly

restricted in their access to public funding for R&D but at the same time heavily dependent on public money. National policies should be aware of this gap and the problems that might emerge from it.

(6) In general, *public policy has to address technology transfer from increasingly important fields*, e.g. R&D of new materials, to the industrial furnace sector. Although it might not be clear from the very start which innovations in the material sciences could contribute to improvements in furnace technologies, it is vital that such opportunities are not missed.

8.4 Final Conclusion and Outlook

Although challenges remain regarding the assessment procedure suggested and applied in this report, overall, the Fh-ISI research team concludes that the method definitely contributes to achieving much greater clarity during the decision process about

- which area of a new energy technology should receive the focus of present and future research and diffusion policy,
- which bottlenecks exist in the technologies as well as in the technological innovation system and
- which policy instruments can best tackle these bottlenecks.

Therefore, the research team proposes applying the manual developed based on the experience gained in this project (see Annex 9.1) to upcoming decisions about future energy research choices by the BMWi or the PtJ and eventually outside support given the demand on interdisciplinary project teams and competences in particular fields such as patent analysis or bibliometrics.

8.4.1 Outlook

This new concept provides both experienced and younger energy research administrators with a quantified and transparent set of indicators allowing additional insights into the innovation system, the obvious bottlenecks and the next actions to be taken and makes their arguments more convincing and open for a rational discussion.

The positive responses of the workshop participants varied in their intensity, but were encouraging enough for the concept to be pursued with passion and enthusiasm at those levels of technology where the additional efforts are more than offset by the benefits of specific indications for further research and policy action and of much greater transparency regarding technical or economic bottlenecks and stakeholder opinions. There is a good chance that the manual derived from the four case studies and the international workshop (see Annex) will contribute to improving the efficiency of energy research and diffusion policy in Germany (and Europe?) in the near future.

The ISI research team suggests a next step to further improve and apply the new concept and to contribute to concrete R&D and innovation policy decision making in the energy field. This next step should close an analytical gap that had to be left open due to the obvious financial and time constraints of the project, i.e. the in-depth comparative analysis of two or three competing technologies. Only if competing technologies have been analysed on an equal footing can our concept fully prove its value in contributing to concrete policy decisions. Given the current activities regarding the design of a new energy programme, there may be a good chance to apply the methodology to some technological innovations such as the fuel cell or carbon capture technologies.

As the proposed methodology is generic in nature, it can also be applied (with some adaptation) to other important fields of new technologies such as health and medical treatment, food processing, or new materials.

Chapter 9
Annex

9.1 Manual

9.1.1 Objective and Outline of the Manual

The objective of the manual is to better structure the considerations involved when deciding whether to promote project-oriented programmes of a field of energy research or when selecting support instruments for market introduction and diffusion on the part of the responsible federal ministry, project sponsor, or large enterprise; to have addressed important issues and to be able to present the considerations interpersonally and transparently.

The manual can be divided into three steps, the second of which is the most complex:

(1) *Market contextualization:* to start with, the technology regarded is assigned to a specific characteristic-category using predefined criteria; this is done in order to conduct certain analysis steps in the subsequent three-pillared analysis with particular care or specific coverage (see Sect. 9.1.2).
(2) *Three-pillared analysis:* this is followed by the three-pillared analysis proposed by the Fh-ISI which is usually the most time consuming activity (see Sect. 9.1.3). The main task of these three analytical steps is the targeted identification of bottlenecks with regard to the technology, the markets or the system of actors and of factors which could especially promote or hinder the further steps along the technology cycle.
(3) *Policy recommendations:* finally, those measures supporting R&D, innovation and markets are identified and evaluated which are seen as essential, according to the position in the technology cycle, to removing the bottlenecks or obstacles identified and to aiding supportive factors.

E. Jochem (ed.), *Improving the Efficiency of R&D and the Market Diffusion of Energy Technologies*, DOI 10.1007/978-3-7908-2154-3_9, © Physica-Verlag Heidelberg 2009

9.1.2 *Characterization of the Technology to Be Examined – Market Contextualization*

First of all, an attempt is made to classify each of the technologies examined using basic criteria which are of considerable importance for the rest of the analysis and significantly determine the priorities set in the subsequent analysis. These characteristics include:

- the regionality of the markets (regional in Europe/global),
- the predominant market form of the technology production and utilization (polypoly/oligopoly of manufacturers or variety of buyers) and
- the intensity of competition due to alternatives to the technology in question.

CO_2 separation technologies in large fossil-fuelled installations (more likely to be made-to-order) such as e.g. in coal-fired power stations, or fuel cells (only competitive as mass products) are cited as examples for *global markets*. In contrast, passive houses represent a technology so far restricted to German-speaking countries and Scandinavia. In the first case there are many global indicators which can be surveyed to obtain an accurate picture of the innovation systems and an appropriate degree of comprehensiveness of indicators, whereas the studies of a regionally limited technology development have to comply with this regional restriction in many indicators (and may be limited as a result).

The *market form* for new technologies which do not require large infrastructures and do not necessarily have to be developed by very large enterprises is frequently polypolistic to start with but then tends to become more oligopolistic with increasing maturity of the technology (e.g. electricity generation, refineries, heat pumps, boilers). There are numerous reasons for this trend, but they are often linked with effects of economies of scale due to experience and the availability of capital to realize mass production. The market form at the point of introduction to the market has an influence on the policy measures selected at almost all stages of the technology cycle (certainly from the disillusionment phase onwards).

The *competitiveness intensity of an alternative technology* – whether this be a traditional one or one developing at the same time with an as yet unresolved outcome of the technological competitiveness – has often not been sufficiently observed (e.g. heat pumps and conventional boilers, the PEM fuel cell and the still improvable Otto engine). If the technological competition is obviously (or could become) an open race, then these technological alternatives should be subjected to a similarly differentiated analysis in order to correctly estimate their ability to penetrate the market or parallel markets.

9.1.3 *Analysis and Evaluation Steps of the Three Pillars*

9.1.3.1 General Principles of the Analysis

To start with, several general rules of thumb are cited which should be adhered to in order to arrive at qualitatively good results as efficiently as possible.

- Try to form an idea early on of *the position of the technology regarded in the technology cycle* because, in certain cases, this could reduce the amount of research necessary for several indicators. However, you should remain open to the possibility of a different positioning in the technology cycle during the further research.
- "Methodological harmony": Try to work on all *three areas of analysis simultaneously* if possible; this will help you "cross pollinate" analysis information and results and will usually accelerate the analytical process by a quicker recognition of core issues and fast feedbacks (e. g suitable interview partners, suitable publication journals or patent classes).
- *You should have in-depth knowledge and know-how in all three analysis areas:* i.e. technological and energy-economic expertise, indicators (patent and bibliometric skills) and experience in analysing innovation systems. If you do not have these available in your team, refer to external experts.
- Pursue the advantage of this structured analysis method as a clear goal*: early perception of the bottlenecks of the technology regarded* and their elements and *assessment of the alternative technology*.
- You must make a *neutral analysis of the competitive technology(ies)*; this rival frequently sets minimum goals for technical features and costs of the new technology. In your first assessment, do not forget the potentials for technology progress and cost reduction, otherwise disappointments are likely (e.g. the competitiveness of the block heating station technology compared with the PEM fuel cell or the microturbine compared with the MFC or SOFC currently underestimated to a large extent).
- The results achieved should be formulated in a *transparent, repeatable procedure*.
- Regularly review the *interim results achieved*, (e. g. in patent classification, publication data, citation analyses and estimates of the cost trend).
- *Adapt your search strategy and the analysis to any difficulties which arise,* e.g. more interviews than originally planned.
- Identify *as quickly as possible the key persons of the innovation system* (e. g. via pre-search interviews or patent and bibliometric analyses).
- Consider that the *effort invested in quantifying and searching for indicators and in interviews has to be worth it* (do not try art for art's sake) and do not cause unnecessary delays to the application and approval process or the administrative process for the policy instruments to be implemented.
- Bear in mind that your team of analysts and their interview partners should use the *same terminology* (by using a glossary with clear definitions of e. g. pilot plant) in order to avoid misunderstandings.

9.1.3.2 The Three Methods of Analysis

Energy Technology and Economic Evaluation

- First estimate the efficiency and emission reduction effects the new technology is capable of achieving with a time horizon of 20–30 years. From this derive an estimation of the potential economies of scale of the new technology based on the assumed market penetration for this period (e. g. from experience curves).

- Note that the traditional technologies or parallel new technology rivals set technical and cost benchmarks which the new technology has to match; if this appears difficult to achieve, make an intensive search for the bottlenecks determining the competition.
- With regard to the question of cost competition, selecting the coefficients of experience curves is an extremely critical point; the procedure is not well developed and there is a great danger of incorrectly assessing future costs as a result of the frequently unclear development of the accumulated production amount.
- Take care also with the assumptions of future energy prices which – often due to vested interests – can lead to serious misjudgements; you should see yourself as an open and level-headed risk manager, avoid or terminate technological R&D lock-in situations by revealing the critical points in good time, since the later this occurs, the more difficult it becomes politically to find a way out of the lock-in situation.

Analysis of the Technology Cycle

- It is vitally important to define the relevant key terms and categories for the patent and publication analyses.
- Avoid a mechanistic application of the indicators at all costs; alternative technologies can have different aggregation levels so that a systematic comparison has to be made (e.g. stationary fuel cells versus separate heat and power generation).
- Note that *each technology has its own specific databases* with regard to the patent classes and the citation sources so that, e. g. the SCI has to be supplemented by specialist publication data. Note also that new technologies may not be contained as such in patent classification and new cross-cutting technologies for the technological progress of traditional technologies may (only) be identifiable by interviews.
- Where new technologies with dual and multiple uses are concerned (e.g. materials, sensors/control, chemo-physical technologies), you should reckon with fuzzy data which is difficult to assign. Address this problem directly during the interviews for a suitable search and analysis strategy to discover the correct patent classes and citation data. Use search words in the patent and citation analyses in order to identify the indicators of the technology application wanted.
- It may not always be easy to identify the position in the technology cycle, this can often only be done iteratively; individual elements of the technology may be in different technology phases (e. g. the rotor of a large wind turbine compared to its generator).
- On account of groups in the innovation system having different estimates, two interest-related positions may result in the technology cycle.

Analysis of the Innovation System

- The first critical and *quality-determining step is the identification of the actor constellations in the technological innovation system and, based on this, the selection of the interview partners*; prior knowledge of the technology and well-known actors can help in the selection as well as the results of patent and bibliometric analyses.
- It is of vital importance here to *cover all the important components of the technology regarded.* Act on the assumption of an ideal technology innovation system and always remember that the significance of components in the innovation system depends on the specificities of the technology and their institutional and organisational localisation (see Point 1 above).
- Note that you need *three waves of interviews*: one from a bird's eye perspective to identify the most important technology elements, one wave of intensive interviews with group-specific discussion guidelines and one to help interpret the indicators of the technology cycle analysis. Try therefore to find *representative* interview partners for the respective group in order to obtain representative answers from the interviews and at the same time make sure to include interview partners with a broad view of the innovation system and the technology.
- *Visualise the respective innovation system*; this is important due to the complexity of actor relationships and to show competitive networks, it also facilitates bottleneck identification and is helpful for projective analyses; this process requires frequent revisions which are easy to do with the right software.
- Bear in mind that *possible interview partners* vary in their *readiness to participate in interviews*, potential interviewees tend to be more open if the technology is still in the research phase.
- In the interviews, do not just focus on questions concerning the development of markets and the technology itself (including bottlenecks), but also on actor constellations and interactions within the innovation system itself.

9.1.3.3 Technical and Cost Targets

- Derive a *catalogue of objectives with regard to the technical and cost targets* for the technology examined from the analysis of competition with technology rivals or from the technical and economies of scale potential.
- Note that the rival technology also has a development potential and that you may have to assume a "moving target".
- The difference between the derived targets and today's technical and cost characteristics of the technology regarded will help you to identify bottlenecks and obstacles.

9.1.3.4 Current Bottlenecks and Market Obstacles

- The *positive identification of technology, cost and system bottlenecks and market obstacles* is the key to a quick and targeted identification of policy measures.
- If there are differing estimations of the position in the technology cycle, the bottlenecks identified should be examined in more detail; perhaps they often have the same localization but their significance is assessed differently.
- Focus your analysis on the bottlenecks and obstacles of the respective position in order to be able to derive the policy measures to be considered in a convincing way.

9.1.4 Policy Measures

Derive the policy measures from the insights gained in the preceding steps.

- The conceivable policy measures result from the interaction of identified bottlenecks in the innovation system ("systemic failures") and in the technology on the one hand and the position in the technology cycle on the other. Keep in mind the characteristics of regionality, market form and competitiveness intensity.
- You can refer to the policy matrix developed to identify possible policy instruments. Bear in mind that the realistic implementation of suitable measures has to be seen against the actual political background and existing policy paths, use the results of the systematic analysis as a lever where necessary in order to break down traditional but counterproductive structures and routines.

You might have to stop the support if you are not convinced that the new technology is technically-economically superior.

9.2 Verfahrensanleitung (Manual in German)

9.2.1 Entwurf der Struktur der Anleitung Ziel der Anleitung und Überblick

Das Ziel der Anleitung ist, die Überlegungen zur Förderentscheidung bei projektorientierten Programmen eines Gebietes der Energieforschung oder der Wahl von Forderinstrumenten für die Markteinführung und -diffusion seitens des jeweils zuständigen Bundesministeriums, eines Projektträgers oder eines Grossunternehmens besser zu strukturieren, wichtige Fragen auf alle Fälle angesprochen zu haben und die Überlegungen interpersonell transparent darlegen zu können.

Die Anleitung gliedert sich in drei Schritte, wovon der zweite der aufwendigste Analyseschritt ist.

(1) *Marktkontexturierung*: Zunächst wird der betrachtete Technologiebereich anhand vordefinierter Kriterien einer bestimmten Merkmals-Kategorie zugeordnet; dies dient dazu, bestimmte Analyseschritte in der darauf folgenden dreigliedrigen Analyse mit besonderer Aufmerksamkeit bzw. speziellem Umfang durchzuführen. (vgl. Abschn.1).
(2) *Dreigliedrige Analyse*: Daran schließt sich dann die seitens des Fh-ISI vorgeschlagene drei-gliedrige Analyse an, die in der Regel den größten Zeitaufwand in Anspruch nimmt (vgl. Abschnitt 2). Die zielorientierte Identifikation von Engpässen in Bezug auf die Technologie, die Märkte oder das Akteurssystem und derjenigen Faktoren, welche die weiteren Fortschritte auf dem Technologiezyklus hemmen oder besonders fördern könnten, ist die zentrale Aufgabe dieser drei analytischen Schritte.
(3) *Policy-Empfehlungen:* Schließlich sollen die FuE- oder innovations- und markt-fördernden Maßnahmen identifiziert und bewertet werden, welche als wesentlich angesehen werden, um gemäß der Position im Technologiezyklus identifizierte Engpässe oder Hemmnisse zu beseitigen und fördernde Faktoren zu unterstützen.

Vorauszuschicken ist allerdings, dass die dargestellte Methodik einen erheblichen Analyseaufwand voraussetzt, der bisher in der vorgeschlagenen Form nicht durchgeführt wurde. Inwieweit dieser jeweils zu rechtfertigen ist, hängt von dem jeweiligen Forschungsprogramm und -feld sowie den damit involvierten Forschungsaufwendungen ab. die hier durchgeführten Beispiele sind sicherlich geeignete Kandidaten infolge weit reichender technologischer Entwicklungen und angesichts erheblicher Risiken infolge von Wettbewerbstechnologien. Diese Empfehlung gilt auch nicht nur für die Phasen von F&E, sondern auch für die Entscheidungen der Pilot-Phasen und Nischenmärkte, wo offensichtlich in der Vergangenheit Fehleinschätzungen zu beobachten sind.

Es sollte auch geprüft werden, ob nicht auch seitens der Antragsteller Zusatzinformationen eingefordert werden können, wie z.B. die eigene Lokalisierung der beantragten Entwicklung in den Technologie-Zyklus und entsprechende Patent- und Literaturzitat-Analysen und die daraus ableitbaren Konsequenzen für die beantragten Forschungs- und Entwicklungsarbeiten oder die Zeitgerechtigkeit von Markteinführungssubventionen.

9.2.2 Kennzeichnung der zu untersuchenden Technologie – Marktkontexturierung

Zunächst wird eine Zuordnung der jeweils zu untersuchenden Technologie anhand grundlegender Kriterien versucht, die für die weiteren Analyseschritte von erheblicher Bedeutung sind und die Schwerpunktsetzung der nachfolgenden Analyse erheblich bestimmen. Zu diesen Merkmalen gehören:

- die Regionalität der Märkte (regional in Europa/global),
- die überwiegende Marktform der Technologieproduktion und -verwertung (Polypol/Oligopol bei den Herstellern oder Vielfalt der Abnehmer) und
- die Wettbewerbsintensität durch potentielle technologische Substitute (technologische Alternativen) der betrachteten Technologie

Als Beispiel für *globale Märkte* sind hier die CO_2-Rückhalte-Techniken aus fossil betriebenen Großanlagen (eher Einzelfertigung) wie z.B. bei Kohlekraftwerken oder die Brennstoffzellen (nur wettbewerbsfähig als Massenprodukte) genannt. Dagegen sind die Passivhäuser bisher eher eine auf die deutschsprachigen Länder und den skandinavischen Raum beschränkte Technologie. Im ersteren Fall sind viele Indikatoren weltweit zu erheben, um ein zutreffendes Bild über das Innovationssystem und eine sachgemäße Vollständigkeit der Indikatorik zu erzielen, während sich die Recherchen einer regional abgrenzbaren Technologieentwicklung dieser regionalen Begrenzung bei vielen Indikatoren entsprechen müssen (und begrenzt werden können).

Die *Marktform* ist bei neuen Techniken, die keine großen Infrastrukturen erfordern und nicht zwingend von sehr großen Unternehmen entwickelt werden müssen, .zu Beginn häufiger polypolistisch, aber mit zunehmender Technologiereife oligopolistisch (z. B. Stromerzeugung, Raffinerien, Wärmepumpen, Heizkessel). Die Gründe für diese Tendenz sind zahlreich, sind aber häufig mit Effekten der Kostendegression durch große Erfahrung und durch Kapitalverfügbarkeit zur Realisierung von Massenfertigung verknüpft. Die Marktform bei Einführung der Technologie in den Markt hat Einfluss auf die Wahl der Policy-Maßnahmen fast allen Stufen des Technikzyklus (sicherlich ab der Desillusionsphase).

Die *Wettbewerbsintensität einer alternativen Technik* – sei sie die bisher tradierte Technologie oder eine parallel sich entwicklende Technik mit offenem Ausgang des technischen Wettbewerbs – wurde häufig nicht hinreichend beobachtet (z.B. die Wärmepumpe und der konventionelle Heizkessel, die PEM-Brennstoffzelle und der noch weiter verbesserungsfähige Ottomotor). Wenn der technische Wettbewerb offensichtlich ein offenes Rennen sein (oder werden) könnte, dann muss diese technische Alternative einer ähnlich differenzierten Analyse zu richtigen Einschätzung ihrer Marktdurchsetzungsfähigkeit oder von Parallelmärkten durchgeführt werden.

9.2.3 Analyse und –Bewertungsschritte der drei methodischen Säulen

9.2.3.1 Allgemeine Prinzipien der Analyse

Zunächst seien einige allgemeine Erfahrungsregeln genannt, die unbedingt eingehalten werden sollten, um möglichst effizient zu qualitativ guten Ergebnissen zu kommen.

- Versuchen Sie, sich relativ früh, eine erste Vorstellung zur *Stellung der betrachteten Technologie im Technologiezyklus zu machen*, weil dies in bestimmten Fällen den Rechercheaufwand für einige Indikatoren vermindern könnte. Allerdings bleiben Sie immer offen für eine andere Positionierung im Technikzyklus bei den weiteren Recherchen.
- "Methodischer Gleichklang": Versuchen sie, möglichst *alle drei Analysebereiche simultan bearbeiten*; hierdurch erzielen Sie: gegenseitige Befruchtung durch Analyseinformationen und -ergebnisse sowie meist eine Beschleunigung des Analyseprozesses durch schnellere Erkenntnis über Kernfragen und schnelle Rückkopplungen (z. B geeignete Interviewpartner, geeignete Publikationsjournale oder Patentklassen).
- *Halten Sie tiefe Kenntnisse und Erfahrungswissen in allen drei Analysebereichen vor*: d.h. technologische und energiewirtschaftliche Kompetenzen, Indikatorik (Patent- und Bibliometrie-Kompetenzen) sowie Erfahrungen zur Innovationssystem-Analyse. Wenn Sie darüber in Ihrem Team nicht verfügen, zeihen Sie externe Kompetenz hinzu.
- Verfolgen sie den Vorteil dieser strukturierten Analysemethodik als klares Ziel: *Früherkenntnis der Engpässe der betrachteten Technik* und ihrer Elemente und *Einschätzung der Alternativtechnik.*
- Sie müssen eine *neutrale Analyse über die Wettbewerbs-Technik(en)* der untersuchten Technologie machen; diese setzt (oder setzen) häufig die Mindestziele für technische Eigenschaften und Kosten der neuen Technologie. Vergessen Sie bei dieser ersten Einschätzung nicht deren technische Fortschrittspotentiale und deren Kostenreduktionspotentiale); andernfalls ist die Wahrscheinlichkeit von Enttäuschungen hoch (z.B. wird die Wettbewerbsfähigkeit der BHKW-Technologie gegenüber der PEM-BZ oder der Mikroturbine gegenüber der MFC oder SOFC derzeit weitgehend unterschätzt).
- Halten Sie die Erarbeitung der erzielten Ergebnisse in einem *transparenten, wiederholbaren Verfahren.*
- Reflektieren Sie regelmäßig die *erzielten Zwischenergebnisse*, (z. B. bei Patentklassifikation, Publikationsdateien, bei Zitieranalysen und Schätzungen der Kostenentwicklung)
- *Adaptieren Sie Ihre Suchstrategie und den Analyseablauf* an die aufkommenden Schwierigkeiten, z. B. zunächst mehr Interviews als geplant.
- Identifizieren sie so schnell wie *mögliche die Schlüsselpersonen des Innovationssystems*, (z. B. durch Suchinterviews oder Patent- und bibliometrische Analysen)
- Bedenken Sie, dass der *Quantifizierungs- und Suchaufwand für die Indikatoren und die Interviews sich lohnen muss* (nicht l'art pour l'art versuchen) und dass Sie durch die Analyse den Antrags- und Genehmigungsprozess oder den Verwaltungsprozess für die in Kraft zu setzenden Policiy-Instrumente nicht unnötig verzögern.
- Bedenken Sie, dass Ihr Analytikerteam und Ihre Interviewpartner *die gleiche Terminologie* (z. B. durch ein Glossary mit klaren Definitionen z.B. zur Pilotanlage) nutzen, um Missverständnisse zu vermeiden.

9.2.3.2 Die drei Analysemethoden

Energietechnische und –wirtschaftliche Bewertung

- Schätzen Sie zunächst die durch die neue Technologie erzielbaren Effizienz- und Emissionsminderungs-Effekte mit einem Zeithorizont von 20 bis 30 Jahren. Daraus ableitbar schätzen Sie mit der unterstellten Marktdurchdringung für diesen Zeitraum die Kostendegressionspotentiale der neuen Technologie (z.B. durch Erfahrungskurven).
- Beachten Sie, dass die traditionellen Techniken oder parallel neue technische Wettbewerber technische und Kosten-Benchmarks setzen, die die neue Technologie leisten muss; wenn dies schwer zu erreichen scheint, schauen sie intensiv nach den Engpässen, die den Wettbewerb bestimmen.
- Bei der Frage des Kostenwettbewerbs ist die Wahl der Koeffizienten der Erfahrungskurven ein extrem kritischer Punkt; das Verfahren ist wenig entwickelt und die Gefahr der Fehlbewertung zukünftiger Kosten infolge der häufig unklaren Entwicklung der kumulierten Produktionsmengen ist groß.
- Achten Sie ebenfalls auf die Annahmen zu den zukünftigen Energiepreisen, die – häufig interessengeleitet – zu erheblichen Fehleinschätzungen führen können; verstehen Sie sich als offene und nüchterne Risikomanager, vermeiden bzw. beendigen Sie technologische FuE-lock in-Situationen durch rechtzeitiges Offenlegen der kritischen Punkte, denn je später dies erfolgt, desto schwieriger ist es politisch, aus der Lock-in Situation wieder herauszufinden.

Technologiezyklus-Analyse

- Legen Sie größten Wert in die Definition von passenden Schlüsselbegriffen und Kategorien für die Patent- und Publikationsanalysen.
- Vermeiden Sie unbedingt eine mechanistische Anwendung der Indikatorik; technische Alternativen können unterschiedliche Aggregationsniveaus haben, so dass der Vergleich systemisch erfolgen muss (z.B. stationäre Brennstoffzelle versus separate Wärme- und Stromerzeugung).
- Beachten Sie, dass *jede Technologie ihre speziellen Datenbanken* bzgl. der Patentklassen und der Zitierquellen hat; so muss z. B. der SCI (Scientific Citation Index) immer ergänzt werden durch fachspezifische Publikationsdateien. Beachten sie auch, dass neue Techniken in Patentklassifizierung als solche nicht enthalten sein mögen und neue Querschnittstechniken für den technischen Fortschritt traditioneller Technologien (nur) via Interviews herauszufinden sein dürfte.
- Bei dual und multiple einsetzbaren neuen Techniken (z.B. Werkstoffe, Sensorik/Regelung, chemo-physikalische Techniken) haben Sie mit unscharfer, schwer zurechenbarer Information der Indikatoren zu rechnen. Sprechen Sie dieses Problem unmittelbar bei den Interviews wegen einer geeigneten Such- und Analysestrategie an, um die richtigen Patentenklassen und Zitierdateien zu erfahren. Setzen Sie bei den Patent- und Zitieranalysen Suchwörter ein, um die Indikatoren der gesuchten technischen Anwendung herauszufinden.
- Die Identifikation der Position im Zyklus einer Technologie mag nicht immer leicht sein, sondern lässt sich häufig nur iterativ klären; auch mögen einzelne

Elements der Technologie sich in einer unterschiedlichen Technologiephase befinden (z.B. der Rotor einer Groß-Wirndturbine im Vergleich zu ihrem Generator).

- Wegen unterschiedlicher Einschätzung von Gruppen im Innovationssystem können interessenbedingt auch zwei Positionen im Technologiezyklus ein Ergebnis sein.

Innovationssystem-Analyse

- Der erste kritische und *qualitäts-bestimmende Schritt ist die Bestimmung der Akteurskonstellationen im technologischen Innovationssystem und darauf aufbauend die Auswahl der Interviewpartner;* bei der Auswahl helfen Vorkenntnisse der Technologie und bekannter Akteure sowie die Ergebnisse der Patent- und biblometrischen Analysen.
- Von erheblicher Bedeutung ist hier die *Erfassung aller wichtigen Komponenten der betrachteten Technik.* Gehen Sie hierbei von einem idealtypischen technologischen Innovationssystem aus und bedenken sie immer, dass die Bedeutung von Komponenten im Innovationssystem auch von den Spezifika der Technologie und ihrer Verortung (siehe Punkt 1 oben) abhängt.
- Beachten Sie, dass Sie *drei Interviewwellen* brauchen: eine aus der Vogelperspektive zur Identifikation der wichtigsten Technologieelemente, eine Welle Intensiv-Interviews mit gruppenspezifischen Gesprächsleitfäden und eine zur Hilfestellung bei der Interpretation der Indikatorik der Technikzyklus-Analyse. Versuchen Sie deswegen, *repräsentative* Interviewpartner für die jeweiligen Gruppen zu finden, um repräsentative Antworten aus den Interviews zu erzielen und achten sie gleichzeitig darauf, dass sie auch Interviewpartner mit einem breiten Überblick über das Innovationssystem und die Technologie einbeziehen.
- *Visualisieren Sie das jeweilige Innovationssystem*; dies ist wichtig wegen der Komplexität der Akteursbeziehungen und zeigt Wettbewerber-Netzwerke, es erleichtert Engpass-Identifikation, und ist auch für projektive Analysen hilfreich; hierbei sind häufige Revisionen erforderlich, die mittels geeigneter Software leicht durchzuführen sind.
- Beachten Sie, dass die *Bereitschaft von möglichen Interviewpartnern zu einem Interview* unterschiedlich ist, in der Tendenz sind die Gesprächspartner offener, wen sich die Technologie noch in den Forschungsphasen befindet.
- Legen Sie den inhaltlichen Schwerpunkt der Interviews nicht nur auf Fragen der Entwicklung von Märkten und der Technologie selbst (inklusive Bottlenecks), sondern auch auf die Akteurskonstellationen und Interaktionen im Innovationssystem selbst.

9.2.3.3 Technische und Kosten-Ziele

- Leiten Sie aus der Wettbewerbsanalyse mit dem technischen Wettbewerber der untersuchten Technologie oder aus dem technischen und Kostendegressionspotential einen *Zielkatalog bzgl. technischer und Kostenzielen* für die jeweils betrachtete Technologie ab.

- Beachten sie dabei, dass auch die Wettbewerbs-Technologie ein Entwicklungspotential hat und Sie vielleicht von einem "moving target" ausgehen müssen.
- Die Differenz zwischen den abgeleiteten Zielen und den Werten der heutigen technischen und kostenseitigen Merkmale der betrachteten Technologie hilft Ihnen bei der Identifikation von Engpässen und Hemmnissen.

9.2.3.4 Derzeitige Engpässe und Markthemmnisse

- Die *eindeutige Identifikation von technologischen, Kosten- und systemischen Engpässen und Markthemmnissen* ist der Schlüssel zu einer schnellen und zielgerichteten
- Identifikation von Policy-Maßnahmen.
- Wenn es unterschiedliche Einschätzungen zur Position im Technologiezyklus gibt, sollten jeweils die identifizierten Engpässe genauer angeschaut werden; vielleicht sind sie häufig gleich lokalisiert, aber unterschiedlich in ihrer Bedeutung eingeschätzt.
- Fokussieren Sie Ihre Analyse auf die Engpässe und Hemmnisse der jeweiligen Position, um daraus überzeugend die zu erwägenden Policy-Maßnahmen ableiten zu können.

9.2.4 Policy-Maßnahmen

Leiten Sie die Policy-Maßnahmen aus den Erkenntnissen der vorangegangenen Schritte ab.

- Die denkbaren Policy-Maßnahmen ergeben sich aus dem Zusammenspiel von erkannten Engpässen im Innovationssystem ("systemic failures") und in der Technologie einerseits und der Position im Technologiezyklus andererseits. Zu beachten sind hierbei auch die Merkmale der Regionalität, der Marktform und der Wettbewerbsintensität.
- Zur Identifizierung von möglichen Policy-Instrumenten können sie die entwickelte Policy-Matrix hinzu ziehen. Beachten sie, dass die realistische Umsetzung der geeigneten Maßnahmen auch vor dem realen politischen Hintergrund und den existierenden Policy-Pfaden zu sehen ist, benutzen Sie die Ergebnisse der systematischen Analyse als Hebel gegebenenfalls, um tradierte, aber kontraproduktive Strukturen und Routinen aufzubrechen.
- Fahren Sie die Förderung eventuell zurück, wenn Sie von der technisch-ökonomischen Überlegenheit der neuen Technologie nicht überzeugt sind.

Printed in the United Kingdom by
Lightning Source UK Ltd., Milton Keynes
141300UK00006B/46/P

9 783790 821536